Principles of Parenteral Solution Validation

Related Titles:

Volumes in the *Expertise in Pharmaceutical Process Technology* Series

- **Mittal, *How to Integrate Quality by Efficient Design (QbED) in Product Development*, August 2019, 9780128168134**
- **Parikh, *How to Optimize a Fluid Bed Processing Technology*, Jan 2017, 9780128047279**
- **Mittal, *How to Develop Robust Solid Oral Dosage Forms: From Conception to Post-Approval*, Oct 2016, 9780128047316**
- **Ostrove, *How to Validate a Pharmaceutical Process*, June 2016, 9780128041482**
- **Levin, *How to Scale-Up a Wet Granulation End Point Scientifically*, Oct 2015, 9780128035221**

Part of the *Expertise in Pharmaceutical Process Technology* Series

Principles of Parenteral Solution Validation

A Practical Lifecycle Approach

Edited by

Igor Gorsky
Senior Consultant, ConcordiaValsource LLC., Downingtown, PA, United States

Harold S. Baseman
Chief Operating Officer, Valsource Inc, Jupiter, FL, United States

Series Editor

Michael Levin
Milev, LLC Pharmaceutical Technology Consulting, NJ, United States

Academic Press is an imprint of Elsevier
125 London Wall, London EC2Y 5AS, United Kingdom
525 B Street, Suite 1650, San Diego, CA 92101, United States
50 Hampshire Street, 5th Floor, Cambridge, MA 02139, United States
The Boulevard, Langford Lane, Kidlington, Oxford OX5 1GB, United Kingdom

British Library Cataloguing-in-Publication Data
A catalogue record for this book is available from the British Library

Library of Congress Cataloging-in-Publication Data
A catalog record for this book is available from the Library of Congress

ISBN: 978-0-12-809412-9

For Information on all Academic Press publications
visit our website at https://www.elsevier.com/books-and-journals

Publisher: Andre Gerhard Wolff
Acquisition Editor: Erin Hill-Parks
Editorial Project Manager: Tracy Tufaga
Production Project Manager: Paul Prasad Chandramohan
Cover Designer: Vicky Pearson Esser

Typeset by MPS Limited, Chennai, India

DEDICATION

I would like to dedicate this book to my family—my wife Faina, my children Jessica and Joshua Gorsky's, my sister Alina Gorokhovsky, my mother Polina Gorokhovsky, and my beloved late father Iosif Gorokhovsky who unfortunately passed away before this book was published. Without them I wouldn't be able to compile this volume. In addition, I would like to thank my co-editor Hal Baseman and Dr. Mike Long for their help and mentorship. Finally, I would like to thank all contributors for their chapters.

Igor Gorsky

CONTENTS

List of Contributors xiii
Editor Biographies xv
About the Expertise in Pharmaceutical Process Technology Series xvii

Introduction 1
Igor Gorsky
References 7
Further Reading 7

Chapter 1 Process Validation: Design and Planning 9
Harold S. Baseman
Part 1: Background—The Need for Process Understanding 9
Why Is Aseptic Process Validation so Challenging? 12
Process Validation, Process Capability, and Process Control 14
Role of Process Understanding 15
Process Life Cycle Approach 16
Proper Process Design as the Key to Process Performance Assurance: Sterility by Design 17
Using a Line of Sight Approach 17
Defining Process Requirements 20
The Role of Process Design and Planning in Validation: Basis of Design 21
Timely User Requirement Specification 22
Equipment and Facility Qualification 23
Mapping the Process 25
Periodic Assessment and Requalification 27
Aseptic Practices: A Key Element in the Validation of Aseptic Processes 28
First Air Principles 30
References 31

Chapter 2 Aseptic Process Validation: Aseptic Process Simulation Design **33**
Harold S. Baseman
Microbial Contamination Case Study: Sterile Vessel Holding Qualification 35
Aseptic Process Simulation Study Design 38
Aseptic Process Simulations Performance Schedule and Frequency ... 39
Media Fills Run Number 39
Inclusion of Process Steps 41
"Worst-Case" Parameters or Conditions 42
Fill Volume 43
Duration 43
Interventions 46
Intervention Evaluation and Risk Assessment Methods 47
Incubation 50
Growth Promotion Studies 51
Preincubation Inspection and Rejection 52
Postincubation Inspection 53
Acceptance Criteria 53
Filled Unit Accountability 54
Failure Investigation 54
Aborted and Invalid Media Fills 55
Special Considerations 55
Powder Filling 56
Ointment Filling 56
Lyophilized Product Filling 56
Anaerobic Processes 57
Conclusion 57
References 58

Chapter 3 Quality Risk Management of Parenteral Process Validation, Part 1: Fundamentals **61**
Amanda McFarland
Quality Risk Management and Process Validation 61
References 79

Chapter 4 Equipment Cleaning Process **81**
Igor Gorsky
Establishing Limits 82

References....96
Further Reading....96

Chapter 5 Quality Risk Management of Parenteral Process Validation, Part 2: A Risk-Based Quality Management System....97
Lori Richter

Overview....97
Deviation Management....99
Corrective and Preventive Action....102
Change Control....105
Change Control and Corrective and Preventive Action Effectiveness Monitoring....106
Self-Inspection Process....107
Summary....113
References....113
Further Reading....113

Chapter 6 Use of Statistics in Process Validation....115
Igor Gorsky

Overview....115
Use of Statistics in Parenteral Process Validation—Ten Basic Concepts....120
Second Concept—Graph It First!....120
Third Concept—Let DataTalk to You!....121
Fourth Concept—Normality?....122
Fifth Concept—Descriptive Statistics....123
Sixth Concept—Control Charting....123
Seventh and Eight Concepts—Tolerance and Capability....125
Ninth Concept—Hypothesis Testing....130
Tenth Concept—Design of Experiments....131
Conclusion....134
References....135
Further Reading....136

Chapter 7 Process Validation Stage 1: Parenteral Process Design....137
Igor Gorsky

Introduction Into Pharmaceutical Development....137
Master Planning, Organization, and Schedule Planning....139

Risk/Impact Assessment 141
Process/System Design 144
Perform Risk Assessment (Identification of Critical Quality Attributes and Critical Process Parameters) 146
Design of Experiments 151
Developing Control Strategies and Determine Process Design 156
Scale-Up and Technology Transfer 157
Stage 1: Design of Experiments Case Study 162
Final Notes About Stage 1 Process Design 172
References 175

Chapter 8 Process Validation Stage 2: Parenteral Process Performance Qualification 177

Harold S. Baseman and Igor Gorsky

General Principles of Stage 2 Life Cycle Approach to Process Validation for Parenteral Products 177
The Line of Sight Approach to Process Understanding and Process Validation 183
Stage 2: Process Qualification 184
Some Points to Consider for the Qualification of Equipment and Systems 189
Critical Utilities 189
Clean Rooms and Classified Areas 190
Component Preparation 192
Product Sterilization and Filtration 194
Filling Equipment and Systems 197
Inspection, Labeling, and Secondary Packaging Systems 198
Lyophilization 198
Material Storage, Handling, and Transport 199
Testing Laboratories 199
Additional Parenteral Processes 200
Cleaning Validation 200
Computer System Validation 200
Terminal Sterilization 200
Container Closure Integrity 200
Periodic Assessment and Requalification 201
Stage 2b—PPQ or Process Performance Qualification 202
Number of Batches 202
Conclusion 205

References 206
Further Reading 207

Chapter 9 Process Validation Stage 3: Continued Process Verification 209
Igor Gorsky

Introduction 209
Definitions Typically Used in Process Validation 213
Determining When Continued Process Verification Starts 216
Legacy Systems Versus New Systems 219
Continued Process Verification Strategy and Enhanced Sampling ... 221
Maintenance of Validation and Change Control, and Periodic Assessment 225
Develop Control Rules for Continued Process Verification 226
Develop Strategies for Continued Process Verification 228
Examples of Case Study Evaluations for Parenteral Products 230
Conclusion 232
References 232

Chapter 10 Preuse/Poststerilization Integrity Testing of Sterilizing Grade Filter 233
Maik W. Jornitz

Introduction 233
Possible Reasons to Apply Preuse/Poststerilization Integrity Testing 234
The Risks Attached to Preuse/Poststerilization Integrity Testing 236
Recommendations 241
References 242

Chapter 11 Environmental Monitoring 243
Igor Gorsky

References 253
Further Reading 254

Chapter 12 Isolators 255
Chris Smalley

Further Reading 259

Chapter 13 Post Aseptic Fill Sterilization and Lethal Treatment 261
Michael J. Sadowski
Introduction 261
Sterilization and Lethal Treatment 262
Summary 267

Conclusion 269
Index 273

LIST OF CONTRIBUTORS

Harold S. Baseman
Chief Operating Officer, Valsource Inc, Jupiter, FL, United States

Igor Gorsky
Senior Consultant, ConcordiaValsource LLC., Downingtown, PA, United States

Maik W. Jornitz
BioProcess Resources LLC, Manorville, NY, United States

Amanda McFarland
Valsource, Inc., Downingtown, PA, United States

Lori Richter
Senior Consultant, Valsource, Inc., Downingtown, PA, United States

Michael J. Sadowski
Lead Scientist, Sterility Assurance, Baxter Healthcare Corporation, IL, United States

Chris Smalley
ValSource Inc., Downingtown, PA, United States

EDITOR BIOGRAPHIES

Igor Gorsky

Igor Gorsky has been a pharmaceutical, biopharmaceutical, and medical device industry professional for over 36 years. He held multiple positions with increasing responsibility at Alpharma (Actavis), Wyeth (Pfizer), and Shire (Takeda). He worked in production, quality assurance, technical services and validation including aa a Head of Validation of the Global Pharmaceutical Technology at Shire (Takeda). He is currently holding a position of a Senior Consultant at ValSource, LLC. His over the years accomplishments include validation of all the aspects of pharmaceutical, biotechnology, and medical device production and quality management, technical support of multibillion-dollar drug product lines and introduction of new and innovative products onto the market. He published numerous articles and white papers in pharmaceutical professional magazines and textbooks, namely *Pharmaceutical Process Scale-Up Handbook*. In addition, he had been a presenter at PDA Annual Meetings, AASP, Interphex, UBM International Workshops, and other forums. He taught classes on validation technology in United States, India, and Israel. He is also very active with PDA participating in several Task force groups authoring PDA Technical Reports 29 (Points to Consider for Cleaning Validation), 60 (Process Validation), 60-2 (Process Validation Annex 1: OSD and SSD Dosage). He is leading PDA Water Interest Group and was named a PDA Volunteer of the Month in Nov/Dec 2014. He also is one of the authors of *ASTM 3106 Standard Guide for Science- and Risk-Based Cleaning Process Development and Validation* and *ASTM G121 Standard Practice for Preparation of Contaminated Test Coupons for the Evaluation of Cleaning Agents*. He is also a member of ASTM E55, Workgroup 159975 Committee Team drafting a Standard Guide for the Derivation of Health-Based Exposure Limits (HBELs). He holds a BS degree in Mechanical/Electrical Engineering Technology from Rochester Institute of Technology.

Hal Baseman,
Chief Operating Officer,
Valsource Inc

Hal Baseman is chief operating officer and a principal at ValSource LLC and ConcordiaValsource LLC. He has over 40 years of experience in pharmaceutical operations, validation, and regulatory compliance. He has held positions in executive management and technical operations at several drug manufacturing and consulting firms. He has held positions as the Chair of the PDA (Parenteral Drug Association), Board of Directors, the Co-Chair of the PDA Science Advisory Board, the Co-Leader of the PDA Aseptic Processing Points to Consider Task Force, and the Co-Leader of the PDA Process Validation Interest Group, as well as a long-time member of the PDA Training Research Institute faculty. He has been a leader, author, editor, and contributor to numerous technical reports, articles, books, and presentations on subjects related to Quality Risk Management, Validation, and Aseptic Processing. Hal holds an MBA in Management from LaSalle University and a B.S. in Biology from Ursinus College.

About the Expertise in Pharmaceutical Process Technology Series

Numerous books and articles have been published on the subject of pharmaceutical process technology. While most of them cover the subject matter in depth and include detailed descriptions of the processes and associated theories and practices of operations, there seems to be a significant lack of practical guides and "how to" publications.

The *Expertise in Pharmaceutical Process Technology* series is designed to fill this void. It will comprise volumes on specific subjects with case studies and practical advice on how to overcome challenges that the practitioners in various fields of pharmaceutical technology are facing.

Format

- The series volumes will be published under the Elsevier Academic Press imprint in both the printed and electronic versions, with each volume containing approximately 100 – 200 pages. Electronic versions will be in full color, while print books will be in black and white.

Subject Matter

- The series will be a collection of hands-on practical guides for practitioners with numerous case studies and step-by-step instructions for proper procedures and problem solving. Each topic should start with a brief overview of the subject matter, exposé, and practical solutions of the most common problems with some case studies, and a lot of common sense (proven scientific rather than empirical practices).
- The series will try to avoid theoretical aspects of the subject matter and limit scientific/mathematical exposé (e.g., modeling, finite elements computations, academic studies, review of publications,

theoretical aspects of process physics or chemistry) unless absolutely vital for understanding or justification of practical approach as advocated by the volume author. At best it will combine both the practical ("how to") and scientific ("why") approach, based on *practically proven* solid theory—model—measurements. The main focus will be to ensure that a practitioner can use the recommended step-by-step approach to improve the results of his/her daily activities.

Target Audience

- Primary audience will be the pharmaceutical personnel, from low level R&D and production technicians to team leaders and department heads. Some topics will also definitely be of interest to people working in nutraceutical and generic manufacturing companies. The series will also be useful for academia and regulatory agencies.
- Each book in the series will target a specific audience, and since the format will be short, the price will be affordable not only for major pharmaceutical libraries but also for thousands of practitioners.

Welcome to the brave new world of practical guides to pharmaceutical technology!

Michael Levin
Series Editor
Milev, LLC Pharmaceutical Technology Consulting, NJ, United States

Introduction

Igor Gorsky
Senior Consultant, ConcordiaValsource LLC., Downingtown, PA, United States

The term *parenteral* pertains to those medicinal preparations that are delivered via the route of injection through one or several layers of skin tissue. This word, derived from the Greek words *para* and *etheron*, which means "outside of the intestine." It is typically referred to those dosage forms that are administered by routes other than the topical and enteral routes. Because the administration of injectables, by definition, requires circumventing the highly protective barriers of the human body, the skin, and the mucous membranes, the dosage form must achieve an exceptional purity and sterility.[1] This is accomplished by strict adherence to good manufacturing practices (GMP) of which process validation plays a pivotal, if not the most important role.

To understand simplest physical presentation of the parenteral products subject of process validation topics to be described in this volume it should be noted that basic principles employed in the preparation of parenteral products do not vary from those widely used in other sterile or nonsterile liquid preparations. This is important to understand from a perspective of uniformity of these products which must confirm to homogeneity specifications of their bulk preparations to be filled and packaged properly in uniform final filled units.

A reader could ask a question why authors are undertaking a discussion on process validation topic for parenteral products when there is already much amount of literature on a subject. What new thoughts and ideas should audience expect from this book? Process validation subject has been a center of attention of regulators and subsequently parenteral industry practitioners for more than 40 years. The subject first was discussed in the mid-1970s, prior to issuance of the FDA

[1] Gorsky (2005).

Principles of Parenteral Solution Validation. DOI: https://doi.org/10.1016/B978-0-12-809412-9.00018-6

compliance program in 1978 (Loftus and Nash, 1984). The program that was issued in 1978 was published before revised cGMP regulations and had a title "Drug Product Inspections." Remarkably this 1978 program said the following with regards to Process Validation:

"A validated manufacturing process is one which has been proved to do what it purports or is represented to do. The proof of validation is obtained through the *collection and evaluation of data*, preferably, *beginning from the process development phase* and *continuing through into the production phase*. Validation necessarily includes *process qualification* (the *qualification of materials, equipment, systems, buildings, personnel*), but it also includes the control of the entire process *for repeated batches or runs*."[2] Interestingly that subject of process validation has been discussed among practitioners for close to 10 years since this definition was first published. Some examples of these discussions were documented in proceeds of Validation of Manufacturing Process Seminar in Geneva, Switzerland in 1980[3] and Validation Seminar held in Dublin, Ireland in 1982.[4] In the proceeds of Geneva seminar we read the following definition for process validation—"a formal process to demonstrate that a specific product can be reliably manufactured *by designed process*," while Validation Seminar in Dublin in one of its presentations asked a question of "how do you "*establish by systematic means?*" or decide that the process is "*grounded on a sound scientific basis*."[5] It is amazing that five years later when Food and Drug Administration's Center for Drugs and Biologics and Center for Devices and Radiological Health[6] Guideline on General Principles of Process Validation "went on books" in May of 1987 original

[2] Food and Drug Administration Compliance Program No. 7356.002. Compliance programs are updated yearly and published in the FDA Compliance Program Guidance Manual, available from National Technical Information Service (NTIS), U.S. Dept. of Commerce, 5285 Port Royal Road, Springfield, VA (information as of 1984).

[3] European Organization for Quality Control, Section of Quality Control in the Pharmaceutical and Cosmetic Industries and Swiss Association for Pharmaceutical Quality, 4th European Seminar on Quality Control in the Pharmaceutical and Cosmetic Industries, Validation of Pharmaceutical processes, September 25/26, 1980, University of Geneva Reports.

[4] Convention for the Mutual Recognition of Inspections with Respect of the Manufacture of Pharmaceutical Products, Validation, A Seminar held in Dublin, 14 and 17 June, 1982.

[5] Convention for the Mutual Recognition of Inspections with Respect of the Manufacture of Pharmaceutical Products, Validation, A Seminar held in Dublin, 14 and 17 June, 1982. Introduction by R. Baker, p. 24.

[6] These are three separate centers now—The Center for Drug Evaluation and Research, The Center for Biologics Evaluation and Research, and Center for Medical Devices and Radiological Health.

1978–1982 definition lost these original concepts of being a life cycle exercise with a sound scientific design of the process and read simply—"Establishing documented evidence which provides a high degree of assurance that a specific process will consistently produce a product meeting its predetermined specifications and quality attributes." One could probably read in "high degree of assurance" a reference to the statistical basis of process validation concepts as statistical examinations into probabilities of planned and designed events to succeed would always include this "high degree of assurance" as in referenced example from a purely statistical manual (Chatterjee, 2003). However in those days pharmaceutical and biopharmaceutical practitioners were not yet given authority by the majority of the management of their organizations to pursue process validation as a continued learning exercise, but rather as a regulatory necessity mainly concentrating on terms "establishing documented evidence" which meant in their mind a "necessary evil" of performing three process validation consecutive batches and documenting this event. It should be noted that even than problems and deviations that often occurred while producing these three "golden" batches did not alert practitioners, their management or even regulators that there was a problem with an entire concept. Firms were regularly cited by the FDA for lack of robust validated processes while manufacturers were continually having issues when they were introducing new products to the market.

In the 1978 GMPs regulators have embedded scientific risk-based life cycle approach into design, qualification, and continued manufacturing of all pharmaceutical products including parenterals. For instance section §211.100(a) states that, "there shall be written procedures for *production and process control designed to assure* that the drug products have the identity, strength, quality, and purity they purport or are represented to possess...." This requires manufacturers to design a process, including operations and controls, which would produce therapeutic materials that would meet their specified attributes. In addition, this section continues that *sampling and testing of in-process materials and drug products*, requires that control procedures "... be established to monitor the output and *to validate* the performance of those manufacturing processes that may be responsible for causing variability in the characteristics of in-process material and the drug product." This reminds manufacturers that even well-designed processes should include in-process control procedures to assure that the design is on a right path and final product quality is guaranteed. The following section 211.110(b) outlines principles for establishment of in-process specifications affirming that they "... shall be derived from previous acceptable process average and process variability estimates where possible and determined by the application of suitable statistical procedures where appropriate." Therefore suggesting that manufacturers should analyze performance of processes to understand and control batch-to-batch variability. Furthermore cGMP regulations sets requirement on sampling, asserting that they must be representative of produced batch under analysis, This described in section §211.160(b)(3) which says that sampling must *meet specifications and statistical quality control criteria as condition of approval and release*, as well as section §211.165(d) which continues saying that the batch must meet its predetermined specifications (§211.165(a)). Additionally regulators also point attention of the producers to an appropriate control of components and drug product containers and closures which could add to variability of production outcome. In section §211.84 (b) cGMPs specifically requires that "representative samples of each shipment of each lot shall be collected for testing or examination," continuing that "the number of containers to be sampled, and the amount of material to be taken from each container, shall be based upon appropriate criteria such as statistical criteria for component variability, confidence levels, and degree of precision desired, the past quality history of the supplier..."

Finally section §211.180(e) requires that information and data about product quality and manufacturing experience be periodically evaluated to determine need for changes in specifications or manufacturing or control procedures.[7] Therefore pharmaceutical products manufacturers would be establishing an ongoing feedback continually evaluating product quality and process performance which is an essential feature of process maintenance and future optimization, improvement, technology transfer, or introduction of new technologies.

In August of 2002 FDA embarked on the "a significant new initiative, Pharmaceutical Current Good Manufacturing Practices (CGMPs) for the 21st Century, to enhance and modernize the regulation of pharmaceutical manufacturing and product quality — to bring a 21st

[7] McNally (2011).

century focus to this critical FDA responsibility."[8] This initiative was significant for many reasons. Process Validation would be playing a much more prominent role than formal data gathering. It would hopefully help manufacturers to establish sustainable cultures based on

- understanding of risks to manufacturing processes as well as;
- building of knowledge bases about their products and processes.

Specifically Final Report for this initiative informed industry that "FDA has also taken steps to clarify" its "our approach to process validation, which was a subject of the 1996 proposal." FDA further stated that they have published a revised compliance policy guide (CPG) entitled Process Validation Requirements for Drug Products and Active Pharmaceutical Ingredients Subject to Pre-Market Approval (CPG 7132c.08, Sec.490.100) and that they intended to further address the validation aspects of the CGMPs by updating the 1987 Guideline on Process Validation, as announced on March 12, 2004. In this new draft guidance they aspire to "address the relationship between modern quality systems and manufacturing science advances to the conduct of process validation."[9] They hoped for this draft guidance to be released in 2005. However it took additional three years for new draft guidance to be published and then three more years for the final draft to be issued. Meanwhile, FDA, along with European Medicinal Agency and Ministry of Health of Japan in a spirit of globalization introduced International Congress for Harmonization (ICH) which in fact followed similar concepts described in cGMPs for 21st Century initiative and issued drafts three pivotal guidances which became pillars of Risk-Based Lifecycle Approach for Process Validation. These of course were ICH Q8 (Pharmaceutical Development), ICH Q9 (Quality Risk Management), and Q10 (Pharmaceutical Quality System). These three guidances include main concepts of Process Validation Continuum based on scientific design, well understood statistically driven measurement system, risk-based analysis and continued knowledge understanding and feedback. However even after introduction of a number of guidances, publishing of voluminous literature, blogs, presentations, and other materials on a subject considerable number of manufacturers seem to fail establishing effective and efficient process validation programs.

[8] Pharmaceutical cgmps for the 21st century—a risk-based approach final report, Department of Health and Human Services U.S Food and Drug Administration, September 2004.

[9] Pharmaceutical cgmps for the 21st century—a risk-based approach final report, Department of Health and Human Services U.S Food and Drug Administration, September 2004.

Therefore the goal of this book is to present readers with practical methods of implementation of process validation programs for parenteral preparations presenting them with real solutions and logical pragmatic methods for cultivating validation cultures within their organizations. This book should be especially useful in an age of globalization to assist practitioners from all over the world in an assembly of a scientific knowledge building blocks that would help them to design, qualify, manufacture, and maintain robust products and processes providing patients with safe, pure, and effective parenteral products.

This book in addition to overview of a life cycle of the parenteral product processes which include discussions about process design, process qualification, and process maintenance will also touch upon subjects that help specifically for aseptic process manufacturing such as use of statistics, aseptic process simulations, post aseptic sterilization, and others.

We divided our book into three sections to help better understanding of the subject and to assure logical flow of information. First four chapters describe enablers of Process Validation for Parenteral Products, analyses' tools and review of essential simultaneous activities such as cleaning validation.

These chapters include:

Chapter 1: Process Validation: Design and Planning.

Chapter 2: Aseptic Process Validation: Aseptic Process Simulation Design.

Chapter 3: Quality Risk Management of Parenteral Process Validation, Part 1: Fundamentals.

Chapter 4: Equipment Cleaning Process.

Chapter 5: Quality Risk Management of Parenteral Process Validation Part 2: A Risk-Based Quality Management System.

After reading these chapters our reader is introduced to actual Process Validation activities which include three stages:

Chapter 6: Use of Statistics in Process Validation.

Chapter 7: Process Validation Stage 1: Parenteral Process Design.

Chapter 8: Process Validation Stage 2: Parenteral Process Performance Qualification.

The final, third section of the book deals with practical aspects of aseptic manufacturing as well as practical activities that are performed during parenteral production. Prospective reader will be introduced to case studies and practical examples in this section. These chapters are:

Chapter 9: Process Validation Stage 3: Continued Process Verification.

Chapter 10: Preuse/Poststerilization Integrity Testing of Sterilizing Grade Filter.

Chapter 11: Environmental Monitoring.

Chapter 12: Isolators.

Chapter 13: Post Aseptic Fill Sterilization and Lethal Treatment.

Our ultimate goal is to present parenteral drug manufacturers with a volume that helps in their implementation of comprehensive process validation programs and hopefully becomes go to reference on this essential subject. We welcome the reader to proceed into a world of parenteral processes validation.

References

Chatterjee, S.K., 2003. Statistical Thought: A Perspective and History. Oxford University Press, Oxford.

Loftus, B.T., Nash, R.A., 1984. Pharmaceutcal Process Validation. Marcel Dekker, New York.

Further Reading

Barker, R. (1982). Convention for the Mutual Recognition of Inspections with Respect of the Manufacture of Pharmaceutical Products, Validation. *Validation Convention*, (p. 24). Dublin.

European Organization for Quality Control, S. o., 1980. Validation of Pharmaceutical Processes. University of Geneva Reports, Geneva.

FDA, 1984. Compliance Program No. 7356.002. US Department of Commerce, Springfield, VA.

FDA, 2004. Pharmaceutical CGMP for the 21st Century - A Risk-Based Approach Final Report. Department of Health and Human Services, US FDA, Rockville.

Gorsky, I., 2005. In: Levin, Michael (Ed.), Pharmaceutical Process Scale-Up. Informa, New York.

McNally, G., 2001. Process validation a lifecycle approach. FDA-PDA Annual Conference. PDA, San Antonio, TX.

CHAPTER 1

Process Validation: Design and Planning

Harold S. Baseman
Chief Operating Officer, Valsource Inc, Jupiter, FL, United States

"We are too busy right now to plan. We'll plan later..." This was what was told to me by a project manager as we contemplated the construction of a new aseptic processing facility many years ago. It may seem laughable but think about the validity of that statement in today's fast-paced, time to market, resource stretched business client. When one hears of companies rushing to perform media fills prior to the completion of filling system and critical utility qualification studies, user requirement specifications (URS) completed after the process design and installation, validation studies that do not link to process performance, environmental monitoring plans implemented without basis, process changes not communicated to those designing, installing, or operating equipment, focus on perceived compliance requirements without regard to scientific and risk-based process requirements, and so on—is the statement all that unbelievable?

This chapter will present principles and methods for planning and designing the validation of sterile parenteral solution manufacturing operations. Parenteral product manufacturing may also include active ingredients, intermediaries, powders, ointments, gels, semisolids, lyophilized, and combination products. However, this chapter will focus on the validation of sterile liquid drug product manufacturing processes. Sterile liquid parenteral products may be manufactured using aseptic processing, terminal sterilization, or a combination. While this chapter will focus primarily on aseptic processing as the manufacturing method, many of the principles and approaches discussed with regard to aseptic processing is applicable to terminal sterilization.

Part 1: Background—The Need for Process Understanding

Proper planning of the validation process is essential for execution of compliant and effective parenteral (or any) manufacturing process

Principles of Parenteral Solution Validation. DOI: https://doi.org/10.1016/B978-0-12-809412-9.00001-0

validation. To validate a process, one must understand the basis of design for the process and its relationship to product quality and output. Designing a manufacturing process begins with understanding what it is we want to achieve. We are designing an aseptic process to achieve a process which is aseptic.

As per the Parenteral Drug Association (PDA), Aseptic Processing is "handling sterile materials in a controlled environment, in which the air supply, facility, materials, equipment and personnel are regulated to control microbial and particulate contamination to acceptable levels." (PDA, 2011). The FDA states that "in an aseptic process, the drug product, container, and closure are first subjected to sterilization methods separately, as appropriate, and then brought together..." (Guidance for Industry, 2004). They further define asepsis as a state of control attained by using an aseptic work area and performing activities in a manner that precludes microbiological contamination of the exposed sterile product (Guidance for Industry, 2004).

In the context of regulated sterile drug manufacturing, aseptic processing is a means to manufacture sterile product where sterile product and product contact components are brought together in such a way as not to contaminate the product or components with microorganisms, that would render the product no longer sterile. If this is true, then what we are validating is the protection of product from contamination.

Aseptic processing may be a challenging process to validate. This was apparent, in the response and reaction by industry to changes in the United States. Current Good Manufacturing Practices (cGMPs) specific to aseptic process validation. In September 2008, the US FDA announced proposed revisions to the cGMPs, Part 211. This revision included changes to Part 211.113 Control of Microbiological Contamination, that added "aseptic" to the requirement for validation of all sterilization processes.

> *(b) Appropriate written procedures, designed to prevent microbiological contamination of drug products purporting to be sterile, shall be established and followed. Such procedures shall include validation of all aseptic and sterilization processes.*

In the explanation posted in the US Federal Register, the FDA stated "...industry routinely conducted validation studies that substituted microbiological media for the actual product to demonstrate that its aseptic processes were validated. These parts of validation studies are often referred to as media fills. ... this revision clarifies existing practices and serves to harmonize the CGMP requirements with Annex 1 of the EU GMPs, which requires such validation." (underlines added by author)

The preamble stated: "Even before 1987, when the Guideline for Sterile Drug Products Produced by Aseptic Processing was issued, industry routinely conducted validation studies that substituted microbiological media for the actual product to demonstrate that its aseptic processes were validated. These parts of validation studies are often referred to as media fills. (Federal Register/Volume 72, No. 232, page 68066)] could lead the reader to incorrectly conclude that only media fills are required to validate an aseptic process."

In comments submitted to the FDA regarding the proposed change, the PDA indicated concerns that the FDA preamble accompanying the proposed GMP changes might be misinterpreted as encouraging media fills as the sole or primary method for aseptic process validation. PDA's comments stated:

PDA believes that a well-controlled, robust process is required for aseptic processes. A highly defined system of risk evaluation and management, engineering and manufacturing controls, maintenance, quality systems, employee training, written procedures, environmental monitoring, strict adherence to aseptic technique, and minimal personnel intervention, can establish a state of control, ensuring that the aseptically produced product consistently meets its pre-determined specifications and quality attributes. Once the state of control has been established, process simulations (media fills) can be useful in confirming the state of control. So, unless the new preamble can be modified, we recommend this section not be revised from the current regulation. The current regulation, and the accompanying preamble, which states, "The Commissioner believes this paragraph, as written, can apply to both sterile fill process and terminal sterilization process. In both instances, there must be validation of the process used to show that it produces a sterile product" provide sufficient regulatory authority for the agency to assure that firms demonstrate a state of control for aseptic processing.

The concern expressed by the PDA's comments was not that the FDA was requiring validation of aseptic processes. That is reasonable and understood. Their objection was that aseptic processes could not be validated using the same methods and approach as sterilization processes. Further, the language in the preamble might lead companies to believe that the running of replicate media fills was the primary way to validate the aseptic process.

Why Is Aseptic Process Validation so challenging?

The validation of the aseptic process is particularly challenging. Some of the reasons for the difficulty are presented in these four points.

1. *Aseptic process validation means trying to prove a negative.* The objective of many validation studies, such as moist heat of sterilization process, is to prove an action has taken place. The objective of sterilization is elimination of microorganisms on or in an item through physical destruction of those microorganisms. To prove this is happening, we determine or designate a level of microbial contamination. We can determine the amount of energy needed to destroy the level of microbial contamination. We know the amount of heat, under what conditions, and for how long is needed to transfer that energy. We can measure and observe the actions of temperature applied, under pressure, for a period. We can then observe and therefore prove that this series of actions have occurred. Proving through observation that actions have occurred, proves the energy has been transferred, proving that the microorganisms would have been destroyed, proving the item has been sterilized.
   However, in the case of aseptic processing, we are proving that an action has not occurred. The objective of aseptic processing is to protect the sterile entity from becoming contaminated with microorganisms. Proving that something has not happened is more difficult and requires a different approach than proving something has happened.
2. *Aseptic processing involves significant variables.* Aseptic processing is complex. It involves interplay between environment, microorganisms, contamination vectors, complicated manufacturing methods, and human behavior. Human behavior is complex, unpredictable, and variable. Training, monitoring, performance feedback, coaching, supervision, observation, ergonomic process design, barriers, and automation can help reduce variability and the effect of variability. Human performance will always be a variable and processes that depend on human intervention and activity will be inherently variable.
   In addition, microbiological behavior and vectors of contamination are complex, often depending on air currents, bioburden, environmental conditions, and material transport. Traditional validation

methods depend on reliable input. Therefore highly variable human activity and environmental condition-related processes are difficult to validate.

3. *Lack of strong correlations.* Current approaches to aseptic process control do not provide a means to measure sterility assurance in aseptic processing. The correlation between what can be observed and the desired or undesired outcome is relatively poor. Air profile (smoke) studies, differential pressures, air velocity, first air principles, ascending and descending clean room classifications, and aseptic technique are designed to prevent microorganisms from contacting and contaminating sterile surfaces or contents. However, these are poor surrogates for indication of contamination. While their effect is intuitive, we do not know for certain how much of a disruption in these controls is needed to contaminate the contents of a vial.
4. *Detection is difficult.* Nondestructive detection of nonsterility is difficult. Environmental monitoring of clean room air and surfaces is imperfect with significant variability in results and capability to obtain meaningful results. Testing is limited by sample size and sensitivity. Environmental monitoring can be imprecise, inadequate recovery and limited sample size.

Part 211.100 of the GMPs requires "... written procedures for product and process control designed to assure that drug products have identity, strength, quality, and purity they purport or are represented to possess." Assurance of a condition, event, or action can be provided by one of the two means. It can be observed, as might be the case with visual inspection of fill volume levels. If one cannot observe a given condition, event, or action, then it can be predicted. Prediction is the declaration of a condition, event, or action that one cannot observe, based on things one can observe. Validation, in a scientific context, is the prediction that an outcome will occur, based on observation and prediction.

One can assure or prove the performance of actions such as cleaning and disinfection, sterilization, compounding, and filling have occurred, because there is a strong correlation between what one can observe and the desired outcome. Proving that something, in this case contamination during aseptic processing, has not occurred is more difficult. There is no one strong observable element or condition that adequately predicts that contamination has not occurred.

The assurance of aseptic process control, avoiding microbial contamination, must be undertaken holistically by addressing each of the supportive elements to optimize them both individually and collectively to provide the greatest confidence in the overall process.

The 2011 revision to PDA's TR 22, the authors wrote that "... the APS is **one tool** for evaluating the processing steps used to manufacture a sterile product. A holistic approach is used to control aseptic processes. An aseptic process incorporates many systems to assure and control sterility of the materials produced, and each should be validated.

- Product, equipment and component sterilization
- Personnel training and certification of aseptic gowning and aseptic techniques
- Equipment and facility sanitization programs
- Environmental controls: microbial levels, differential pressure, air pattern, velocity, temperature and humidity
- Personnel, material and equipment flows
- Process design"

Validation of the aseptic process is not accomplished by one media fill test or set of tests performed at the end of the process. As is the case with the validation of any complex process, it is accomplished by carefully and scientifically evaluating the design and performance of all process steps that can adversely affect the outcome of the process (PDA, 2011).

Process Validation, Process Capability, and Process Control

To understand how best to validate the aseptic process, it is important to understand the basis and logic inherent in modern process validation risk-based thinking. And to do that one should understand the background, purpose, and objective of process validation.

In 1970–71 outbreaks of *E. cloacae* and *Erwinia* contamination were noted in LVP bottles. This contamination resulted in significant patient injury and some deaths. The contamination was apparently caused by moisture seeping into a space under the screw cap of the bottles during cooling after sterilization. Standard sterility testing did not pick up the contamination, because the cap and septum were removed prior to test. Thus it became apparent that testing and process monitoring was not enough to assure the quality of the product. Product quality assurance, where the attributes of the product could not be adequately observed or tested, needed to be based on proving that the process was capable. The process of proving or providing confidence of the capability and reliability of the process to achieve desired outcomes became known as validation (Agalloco, 1995).

Role of Process Understanding

Over ensuing years, the US FDA compiled the best practices of industry for process validation and in 1987 published its Process Validation Guidance for industry. In the 1987 guidance, process validation was defined as *establishing documented evidence which provides a high degree of assurance that a specific process will consistently produce a product meeting its pre-determined specifications and quality characteristics.* This guidance set many of the principles we used in modern process validation, including facility and equipment qualification, prospective and concurrent validation, requalification, the need for good documentation, as well as the reliance on replicate successful runs to demonstrate process capability. If one can successfully run a process say three times without significant failure, then does that prove it can be run all the time without failure? If the answer is yes, then the process is validated. However, if so, then why do processes occasionally fail after such validation? The answer is that things happen, things change. These things that change are reflected as process variables. More clearly stated, they are unforeseen or inadequately addressed process variables (FDA, 1987).

The key to better process validation, then, would be more effective means to identify and address process variables. Unresolved process variables represent weakness in the process control strategy. The best time to identify these variables and weakness is early in process design. With these concerns in consideration and an understanding that the identification and address of process variables was key, the US FDA announced in 2006 its intention to revise the 1987 Process Validation Guidance. In November 2008, they issued the draft revision and published the final version in January 2011.

This guidance expressed a logical, risk, and scientific thinking-based approach that is well described in a new definition of process validation as "... the collection and evaluation of data, from the process design stage throughout production, which establishes scientific evidence that a process is capable of consistently delivering quality products." In the 2011 guidance, the emphasis shifts from the establishment of documented evidence from qualification and validation studies to the evaluation of data which establishes scientific evidence throughout the process life cycle. The 2011 guidance moves focus from the results of set qualification and validation studies to the work needed to identify and control process variables. Thus the shift in the way confidence in the process is obtained and process quality assured (FDA, 1987).

Process Life Cycle Approach

The 2011 guidance presents a three-stage approach as described in Fig. 1.1. Here Stage 1 includes the design of the process, including the identification of process variables. Stage 2 tests the controls in place to address the variables. Stage 2 also includes traditional facility, equipment, and system qualification. Stage 3 represents ongoing monitoring and evaluation of the process (FDA, 1987).

Note the evolution of quality assurance thinking from reliance on the testing of product (pre-1970s) to reliance on the outcome of qualification and validation tests primarily (1980s through 2010) to the

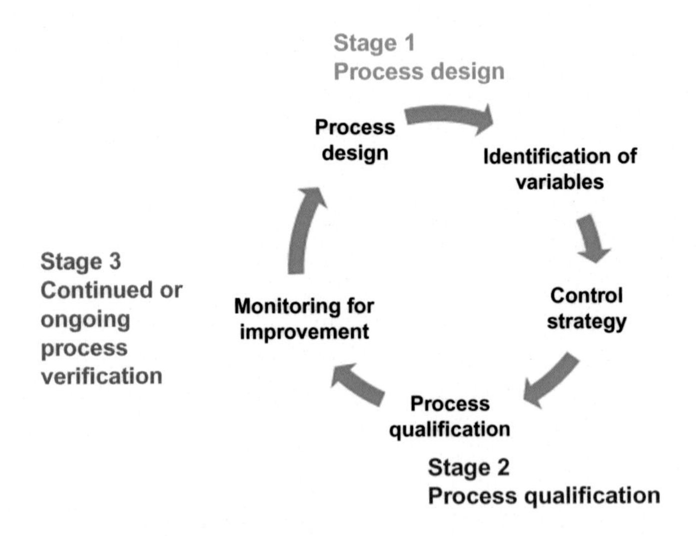

Figure 1.1 Three-stage life cycle approach to process validation.

uncovering of process variables from process development through commercial manufacturing (today). This denotes a shift from qualification and validation as distinct events to validation as a continuous process of knowledge transfer and understanding, where emphasis is on confidence through proper process design rather than solely on testing. This also presents an emphasis on prevention of failure, rather than detection of failure.

Proper Process Design as the Key to Process Performance Assurance: Sterility by Design

An answer to more effective process validation is in the identification of as many process variables as possible, if not all (Agalloco and Akers, 2005). Attaining primary confidence in the process should happen during the design of the process, rather than during final process qualification testing or continued process verification. If one relies too much on Stage 3 continued process verification to provide process control confidence, then if a problem or process weakness is discovered, potentially compromised product may be in the market. If one relies too much on Stage 2, then if a problem or process weakness is discovered, the process may require costly changes, or the company may choose to work with a suboptimal and marginally effective process. Stage 1, process design, is when one should have near complete confidence in the reliable performance and control of the process. It is that confidence that allows the company to commit to the process design. Stages 2 and 3 sampling, testing, and monitoring are designed to confirm decisions made in Stage 1 and ensure that variables were not missed in the initial evaluation and that new variables were not introduced after that evaluation.

Using a Line of Sight Approach

To do that, information must be gleaned from early process development stages. A line of sight (LOS) approach, as noted in Fig. 1.2, can be used.

In a LOS approach, the objective of the qualification or validation is kept in sight so as to speak. All actions are focused on achieving that objective. The objective in this case may be the establishment and maintenance of a critical quality attribute (CQA). All actions taken

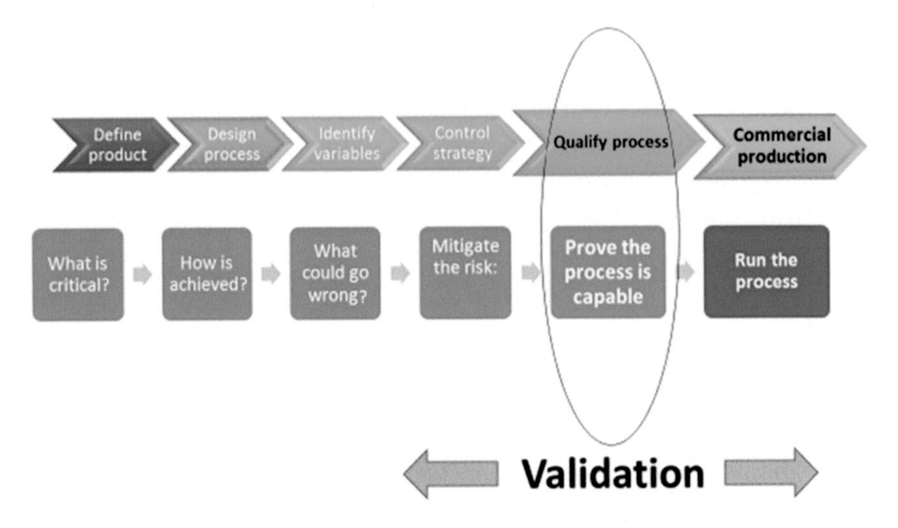

Figure 1.2 Line of sight process control and validation.

during the qualification and validation should be able to be linked to meeting that objective and all objectives should be addressed by one or more qualification or validation actions.

The LOS approach to process control and validation can be described in six progressive steps.

1. In Step 1, the CQAs of the product are identified. These are the conditions and functions that define the product as useful. These might include specifications for strength/potency, safety/sterility, purity, identity, and functionality. Where possible, it is important that these attributes be stated as quantified, measurable criteria, or specifications. This will help in designing process control strategy and actions and in determining process qualification tests and acceptance criteria.
2. Step 2 determines the process steps needed to achieve those quality attributes. Each quality attribute relies on one or more processes to achieve that attribute. For example, sterility might involve clean room, sterilization of components, sterilization of product, sterilization of filler parts, assembly of sterilized filler parts, holding of sterile materials, transport of components, and sealed containers.
3. Step 3 considers the variables that might be inherent in those process steps that could result in loss of, or pose a risk to, those quality attributes. These might include operator error, poor process design,

failure of sterilization processes, nonintegral packaging or container integrity, environmental contamination, etc.
4. Step 4 determines the controls or control strategy needed to mitigate that risk. At the end of this step, the company should have the understanding and confidence in the process and associated controls to finalize the design and install the systems needed to perform the process. These might include gowning qualification, environmental monitoring, clean room disinfection, load pattern and recording of sterilization cycles, design of sterilization circuit, container integrity inspection, and testing.
5. Step 5 develops the means to test if those control measures are effective in addressing process variables and risk. This might be considered where "traditional" IOQ (installation/operational) qualification and process performance qualification validation studies occur. Once Step 5 is complete, the company should have the confidence in the reliability and capability of the process to begin commercial manufacturing. This might include design and installation qualification of clean room, HVAC system, critical utilities, operational qualification of filling lines, and aseptic process simulations.
6. Step 6 occurs during commercial manufacturing. It involves the continued or ongoing acquisition and evaluation of information obtained from performing the process during manufacturing. The step is designed to capture additional information needed to assure that the process is maintained in a qualified state, that additional variables have not been previously uncovered or adequately controlled, and if so, address those variables to improve the process. This might include environmental monitoring, process performance metrics (e.g., CpK, PpK), process failures, investigations, OOS results, sterility failures, and audits.

These steps are aligned with the 2011 FDA guidance, with Steps 1–4 comprising Stage 1 (process design), Step 5 being process qualification, and Step 6 continued process verification. Fig. 1.3 illustrates the relationship between the three-stage life cycle approach and some of the common aseptic process validation activities.

Control strategies and process steps designed to mitigate one risk may add additional risks to other product attributes. The interaction of process steps may pose a risk to the quality of the product. These unintended consequences should be considered during the validation process. For example, the use of a more aggressive disinfectant and sanitization

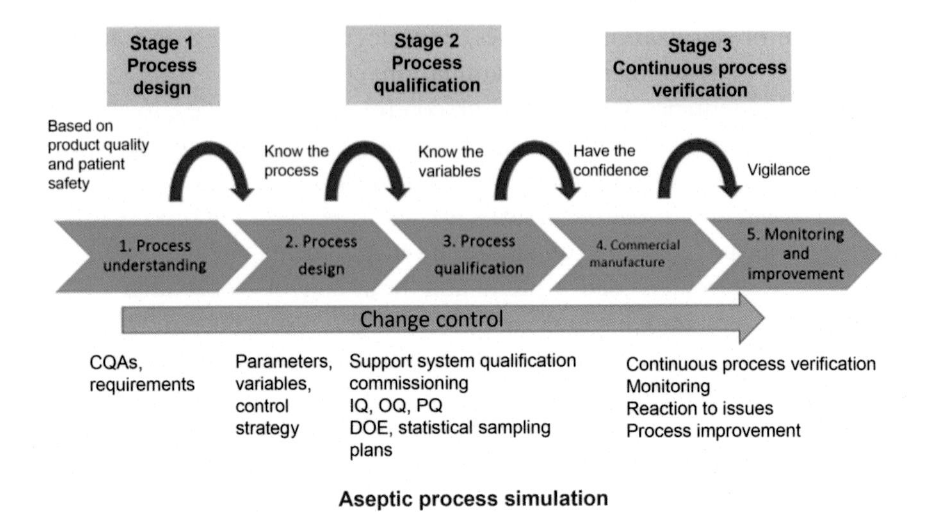

Figure 1.3 Combined process validation sequence.

program may be a control strategy for reducing the risk of environmental contamination, but may also damage filling equipment or pose a risk of chemical contamination to product. In this case, it would be important not only to address the intended benefit of the mitigation action, but also the potential unintended consequence of the action.

Defining Process Requirements

The manufacturing of sterile parenteral drug products will involve several key steps. These steps will depend on the objective of the manufacturing operation, as reflected in desired outcome of the process or the production needs of the operation. This would include type of products and therapies, annual production output required, rate of output, number of products, global regulatory requirements, flexibility and redundancy business strategy, cost strategy, and production timing. This is where the information needed for Column 1 in Table 1.1 can be obtained. This information is often contained in a written quality profile. Quality Target Product Profile is a prospective summary of the quality characteristics of a drug product that ideally will be achieved to ensure the desired quality, considering safety, and efficacy of the drug product. "US FDA Guidance for Industry and Review Staff: Target Product Profile—A Strategic Development Process Tool" [US FDA Guidance for Industry and Review Staff: Target Product

Table 1.1 Line of Sight Process Evaluation (Safety–Sterility)

1	2	3	4	5	6
CQA	**Process Steps**	**Process Variables**	**Control Strategies**	**Qualification Testing**	**Ongoing Verification**
Safety–sterility	Sterilization, container closure integrity, aseptic processing	Inadequate temperature and exposure time, air in system, steam conditions, incorrect or random loading patterns, excessive bioburden, incorrect components poor aseptic and gowning technique	Autoclave design, cycle monitoring/recording, established loading patterns, automated receipt and storage of components, clean component and ingredient storage conditions, properly sized filters, training, first air principles, smoke studies	Autoclave IOQ, temperature mapping, BI studies, air removal studies	Continued process monitoring
					Data analyses from manufactured lots
					Ongoing periodic autoclave qualification

Profile—A Strategic Development Process Tool, CDER, Draft, March 2007 (PDA, 2013) (PDA TR 60)]

Sometimes referred to as a project charter, the high-level objective of the production process should be noted, including information such as:

- annual or periodic production quantities;
- CQAs of the product; and
- patent or approval timing issues.

The Role of Process Design and Planning in Validation: Basis of Design

The basis of design defines the limitations and requirements to meet the objectives of the manufacturing operation and process. These limitations and requirements may be driven by questions related to technical, regulatory, market, cost, safety, business, or other concerns and circumstances. For example:

- Will the product be mass produced?
- Can and will it be sterilized by filtration?
- Can and will it be terminally sterilized?
- Is it sensitive to air, light, or temperature?
- Can it be sheared or damaged through the filling process?
- Will it react to manufacturing and fill system materials of construction, such as metals, stainless steel, and plastics?

- Must it be lyophilized for functionality or stability?
- Does it involve autologous and allogeneic cell collection?
- Does it have special or unusual holding, storage, collection, or transport requirements?
- Is it hazardous, hypoallergenic, or requires containment considerations?
- In what jurisdictions and markets will it be distributed?
- What are the cost limitations?
- Are there unusual stability limitations?
- Will product be produced in existing facility, and if so, then—what are the limitations of that facility?
- Will redundant manufacturing systems be required?

The product requirements, along with contamination control and efficiency strategies will influence technology decisions. As such, the answers to these questions will in part determine the technologies and methods used to manufacture the products, including the use of:

- aseptic or terminal sterilization processes;
- batch or continuous process manufacturing;
- manual fill systems;
- conventional semiautomated filling systems;
- barrier systems such as open and closed RABS (restricted access barrier systems);
- isolators;
- blow fill seal and form fill seal;
- closed vial filling;
- robotics and automated filling systems;
- steam/sterilize in place product filtration and transfer systems;
- PUPSIT (poststerilization, preuse integrity testing) of sterilizing filters; and
- continuous total particulate and/or viable environmental monitoring systems.

The outcome of this evaluation will help provide the information contained in Column 2 of Table 1.1.

Timely User Requirement Specification

EU Annex 15 section 3.2 well states the purpose of the URS (EMA Annex 15; PDA, 2014): "The specification for equipment, facilities, utilities or systems should be defined in a URS and/or a functional specification. The essential elements of quality need to be built in at this stage and any GMP risks mitigated to an acceptable level. The URS should be a point of reference throughout the validation life cycle."

The URS is important to parenteral manufacturing process validation, because it sets the criteria for process performance. The URS is developed once the basic technology strategy is set. The URS presents more specific user needs that can be translated into a manufacturing process design. The URS should be prepared prior to the design of the process and the selection of systems. The URS can then be used to determine the critical aspects of the process, process parameters, and qualification/validation acceptance criteria. Information and design requirements noted in the URS should then be used as the basis of the design qualification to confirm that the critical process requirements are incorporated into the system design. The URS should address process anticipated variables and control strategy to ensure reliable process performance and output meeting production/product requirements. The URS can be used to obtain information needed to complete Columns 2 and 3 of Table 1.1.

The following example CQA's and process steps which they are measuring could be identified using LOS approach as shown in Table 1.1:

- Strength—drug potency: Process Step: Formulation and compounding, raw material handling, weighing, exposure to light
- Purity—absence of foreign materials: Process Step: Cleaning, proper handling and storage, environmental controls, WFI generation
- Identity—correct labeling: Process Step: Labeling, inserts, secondary packaging, decorated glassware, lot coding serialization
- Functionality—container and packaging: Process Step: Human factors insert instructions for use, torque

Equipment and Facility Qualification

The 2011 FDA Process Validation Guidance states that the foundation for process validation is provided in § 211.100(a), further stating that "'[t]here shall be written procedures for production and process control *designed to assure* that the drug products have the identity, strength, quality, and purity they purport or are represented to possess...' This regulation requires manufacturers to design a process, including operations and controls, which results in a product meeting these attributes. In addition, the CGMP regulations require that facilities in which drugs are manufactured be of suitable size, construction, and location to facilitate proper operations (§ 211.42). Equipment must be of appropriate design, adequate size, and suitably located to facilitate operations for its intended use (§ 211.63). Automated, mechanical, and electronic equipment must be calibrated, inspected, or checked according to a written program designed to assure proper performance (§ 211.68)."

Before aseptic process validation studies can be conducted, it is essential that the performance of the systems, equipment, and facilities are reliable, consistent, without unforeseen variation. Systems, equipment, and facilities that support aseptic processes, by establishing, controlling, or maintaining sterile product CQAs, should be qualified. The overall objective of qualification is to eliminate system operation as a variable in the assessment of the effectiveness of the process. Qualification shows that a system used in the process is fit for use; capable of performing the process in a reliable manner.

Typical qualification-related terminology (see Fig. 1.4):

1. DQ or design qualification confirms that system design meets URS (and is compliant with relevant regulatory requirements and expectations).

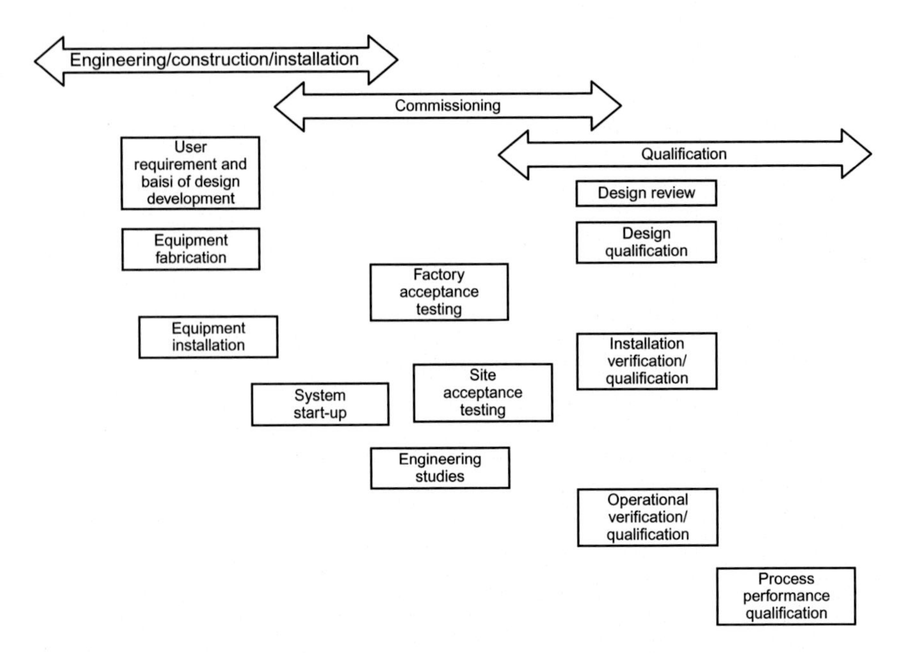

Figure 1.4 System, equipment, and facility qualification sequence.

2. IQ or installation qualification confirms that system has been installed according to design.
3. OQ or operational qualification confirms that system operates to its intended use (or its capability).
4. FAT or factory acceptance testing is work performed at place of fabrication confirming that system is suitable for delivery. Information from FAT may be leveraged or used in support of IQ and OQ.
5. SAT or site acceptance testing is work performed at site of use/installation confirming that system is ready for turnover. Information from SAT may be leveraged or used in support of IQ and OQ.
6. PQ or performance qualification confirms that the system consistently performs the process to expectations.
7. PQ or process qualification is the overall term for Stage 1 of the FDA Process Validation Guidance life cycle approach.
8. PPQ or process performance qualification is the term for that part of Stage 2 of the FDA Process Validation Guidance that involves confirmation batches.
9. CPV or continued process verification, also referred to as ongoing process evaluation in EU annex 15 is the term used to describe the continued collection and evaluation of information throughout the system life cycle, assuring that during routine production the process remains in a state of control: A stage of the Process Lifecycle, after Performance Qualification (FDA Guide) (PDA, 2011) FDA Stage 3 activity.

Mapping the Process

Typical parenteral process might include the following process steps as presented in Fig. 1.5.

Determine the critical aspects of each process step and methods to qualify and test that the control measures in place to ensure process performance work properly and reliably.

What are the critical aspects of this process step, which can adversely affect the quality of the sterile parenteral product?

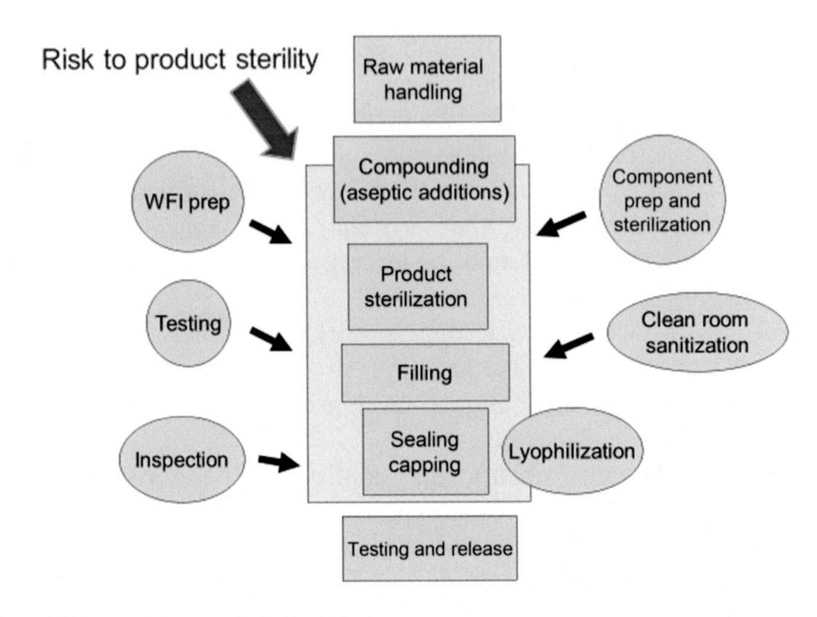

Figure 1.5 Parenteral drug manufacturing steps.

Process Step	Material Handling and Storage
Process variable	Materials can be mixed up during receipt, handling, and storage
Control measure	Identification of materials upon receipt with automatically generated labels and distinctive code
Qualification	IOQ and computer system validation of label generation system and code reader

Process Step	Material Handling and Storage
Process variable	Materials can become contaminated during sampling, resulting in excessive bioburden
Control measure	Material opened in controlled environment sampling station
Qualification	IOQ and controlled environment testing, review of training and procedures for clean handling of material

Process Step	Material Handling and Storage
Process variable	Materials can be exposed to excessive temperature during storage, resulting in loss of potency
Control measure	Material stored in controlled temperature area
Qualification	IOQ and controlled temperature area monitoring

Process Step	Compounding of Product to be Filled
Process variable	Product can be contaminated with prior product, resulting in loss of purity
Control measure	Cleaning in place of compounding tank and lines
Qualification	IQ to determine that product contact surfaces are cleanable, OQ to confirm cleaning process parameters are met, cleaning validation to ensure batch residuals are removed, and controlled environment testing

Process Step	Sterilization of Filler Parts
Process variable	Inadequate moist heat exposure to destroy microbial contamination of parts
Control measure	Autoclave cycle is monitored and recorded, autoclave is fed with proper quality of steam, parts are wrapped and loaded properly
Qualification	IOQ

The information obtained from this type of evaluation can be used to help determine qualification needs as stated in Columns 5 and 6 of Table 1.1.

Periodic Assessment and Requalification

Once a system or piece of equipment has been qualified, it is important that it be maintained in a qualified state. This involves ensuring that the system is well maintained, calibrated, and operationally fit. It also means that additional variables from intrinsic and extrinsic factors do not adversely affect its operation.

Continued process verification involves reviewing the systems on an ongoing basis and may require additional testing to verify control. This shifts the emphasis from periodic/time-driven requalification to periodic assessment. An ongoing periodic assessment schedule can be set based on confidence in function and risk to product quality, including such aspects as system function, system robustness, impact of system function or failure on product quality, system familiarity, system complexity, redundant control systems, monitoring systems, and operational history.

The assessment might consider indicators that show if the system remains in qualified state, including such aspects as maintenance status, work orders completed, change order extent and status, changes in regulatory expectations or requirements, system failures and investigations, system performance monitoring results, timing and results of previous qualification, or other evidence of loss of control.

Based on the results of the periodic review, the system may require actions, additional information and assessment, and additional qualification or requalification. Risk assessments can be used to establish the periodic assessment program, schedules, criteria, and help with decisions on results.

Aseptic Practices: A Key Element in the Validation of Aseptic Processes

Aseptic manufacturing processes include steps that expose surfaces or materials to potential sources of microbiological contamination. These sources may include people, the environment, equipment or facility surfaces, components, parts, other materials, utilities, gases, water, ingredients, etc. The contamination may be relatively easy to monitor and detect, as might be the case with bioburden of materials, difficult as might be the case with environmental contamination, sporadic as might be the case with biofilm, or very difficult as might be the case with human intervention related contamination.

Therefore the protection of sterile product and product contact surfaces becomes very important to the reliable performance of the aseptic process. The protection of these elements is the objective of good aseptic practices. It is that objective that should be captured in the validation approach.

Aseptic practices designed to protect the product from microbiological contamination include things such as:

- clean room design;
- cleanable clean room surfaces;
- clean room equipment design;
- clean room equipment cleanable surfaces;
- positioning of equipment and functions in relation to first air principles;
- clean room HEPA filtered air;
- clean room differential air pressure;
- clean room HVAC temperature and humidity controls;
- clean room air changes;

- design of adjacent clean room spaces and areas;
- work, personnel, equipment, materials, and component flow;
- wrapping, holding, and assembly of sterilized product contact parts;
- wrapping, holding, and transfer of sterilized components and containers;
- holding and transfer of sterilized product;
- product transfer;
- inherent and corrective interventions;
- environmental monitoring practices and techniques;
- environmental monitoring sampling plans;
- clean room personnel supervision and observation;
- clean room personnel rest periods, breaks, fatigue, ergonomic process design;
- clean room gowning;
- sanitization materials entering the clean room;
- attachment of lines, filters, tanks, vessels, hosing, etc.;
- personnel personal hygiene and illness recognition;
- exclusion or restriction of materials, personnel, items and such from clean room;
- deviation communication and handling; and
- overall aseptic technique.

Each of these activities or elements have an impact on protecting and maintaining the sterility of parenteral product. Therefore each should be addressed in the validation program. Of course, there are other parenteral manufacturing process steps that can affect product sterility, including:

- presterilization solution compounding;
- cleaning, sanitization, and sterilization of takes, vessels, filters, and lines;
- cleaning, sterilization, and depyrogenation of components;
- cleaning and sterilization of parts;
- sterilization or filtration of product;
- decontamination of "closed" process environments;
- container closure sealing and integrity; and
- inspection of filled units.

These process steps and actions are essential in a holistic approach to aseptic process validation. Each process should be validated and equipment associated or in support of the process should be qualified. However, because these steps are involving actions are validated in separate studies, this section of the chapter will focus on those process steps and aspects of the process that are designed to protect sterilized product from microbiological contamination and are handled through aseptic practices and techniques.

First Air Principles

Aseptic practices are those process steps and related functions performed in a protected, classified environment, using proper aseptic technique, and adhering to first air principles. First air refers to uninterrupted HEPA filtered air traveling from the source, HEPA filter, to the sterile entity. Observing first air principles involves using good aseptic technique to avoid placing a potentially contaminated object in between that flow of air and the sterile entity (product or product contact surface). Interventions using sterilized objects, such as sterilized forceps, handled correctly, that disrupt first air have a relatively low risk of contaminating the sterile entity. Interventions using nonsterile objects, such as gloved hands, or gowned sleeves, that disrupt first air have a higher risk of contaminating the sterile entity. The risk can be mitigated by designing the airflow so that the air from the HEPA filter flows to the sterile entity and then to the potentially contaminated object, as illustrated in Fig. 1.6. The air should never touch the contaminated object and then flow back to the sterile entity.

Aseptic practices rely, in part, on clean air in the critical area surrounding the sterile entity, directed airflow, good process design, clean

Figure 1.6 First airflow. ivtnetwork.com/gallery/aseptic-core-images, 2009.

surfaces, and proper aseptic technique. The validation of aseptic practices should address those variables that could adversely affect the effectiveness of the aseptic practice. For example, the variables associated with clean area might include capability and integrity of HEPA filter, differential pressure, air flow and velocity and air changes. The variables associated with directed airflow might include unidirectional airflow, air velocity, interference, and turbulence. The variables associated with process design might include ergonomics, equipment design, operator access, and barriers. The variables associated with clean surfaces might include porous surfaces, seams, reaction to cleaning materials, and limited access. The variables associated with aseptic technique might include gowning, training, access, excessive exposure, and obstacles.

Aseptic process simulation or media fills might be a tool for proving confidence that the aseptic process was designed and performed properly. However, there are more effective ways to provide confidence that aseptic practices are in place. Confidence in the aseptic practice components of the aseptic process may be better gained through HEPA efficiency testing, velocimeter readings, smoke studies, design review, disinfection efficacy studies, gown qualification, training, and observation.

References

Agalloco, J., 1995. Validation: an unconventional review and reinvention. PDA J. Pharm. Sci. Technol. 49 (4), 175–179.

Agalloco, J., Akers, J., 2005. Risk analysis for aseptic processing: The Akers–Agalloco method. Pharm. Technol. 29 (11), 74–88.

European Commission, EudraLex Volume 4, EU Guidelines for Good Manufacturing Practice for Medicinal Products for Human and Veterinary Use, March 2015 Annex 15: Qualification and Validation.

FDA, 1987. Guidelines on General Principles of Process Validation. Food and Drug Administration.

Guidance for Industry, Sterile Drug Products Produced by Aseptic Processing—Current Good Manufacturing Practice, U.S. Department of Health and Human Services Food and Drug Administration, Center for Drug Evaluation and Research (CDER), Center for Biologics Evaluation and Research (CBER), Office of Regulatory Affairs (ORA), September 2004, Pharmaceutical CGMPs.

PDA, 2011. Technical Report No. 22 (Revised 2011), Process Simulation for Aseptically Filled Products, PDA Bethesda.

PDA, 2013. Technical Report No. 60 Process Validation: A Lifecycle Approach.

PDA, 2014. Technical Report No. 54-4 Implementation of Quality Risk Management for Pharmaceutical and Biotechnology Manufacturing Operations.

CHAPTER 2

Aseptic Process Validation: Aseptic Process Simulation Design

Harold S. Baseman
Chief Operating Officer, Valsource Inc, Jupiter, FL, United States

Aseptic process simulations, sometimes referred to as media fills, are studies conducted on the aseptic filling process. The process is simulated or run as closely to the actual production procedure as possible. Product is replaced with growth media. The growth media is selected to act as closely to the product properties as practical and still support and indicate relatively low levels of microbial contamination. If there are flaws or weakness in the process capable of resulting in product contamination, then that contamination will be more likely to be discovered in a growth supporting media, than in product, thus exposing the weaknesses and flaws of the process.

As noted in an earlier chapter, one of the reasons aseptic processes are difficult to validate is that it involves trying to prove something has not happened. One of the principles of the scientific method can help with this proof. If one wishes to prove a hypothesis, one can try to disprove it. In failing to do so, an element of proof is achieved. Here the hypothesis is that the aseptic process will result in sterile product or perhaps better say, will not result in microbiological contamination of the sterile product. In other words, the aseptic process is devoid of flaws, weaknesses, and underaddressed variables serious enough to result in microbial contamination. Filling media in place of product is an attempt to challenge the process and find those weaknesses. Conducting multiple media fills will increase the likelihood of uncovering process flaws, weakness, or variables. More simply put, if you can run the process three times without failure, then it must be good.

But is this enough of a challenge? Will only running three media fills provide enough of a challenge to uncover all underaddressed process variables or weaknesses to provide the requisite high degree of

Principles of Parenteral Solution Validation. DOI: https://doi.org/10.1016/B978-0-12-809412-9.00011-3

assurance that the process will result in a prescribed level of product quality? And if so, then could you explain why running three media fills would validate this process?

As stated in an earlier chapter, media fills are but one tool used to validate the aseptic process.

Part of overall process validation approach involves challenging capability of the aseptic process. Good validation practices rely on the recognition of the effectiveness and limitations of the study to meet process control objectives. As stated in PDA Technical Report No. 22, Process Simulation for Aseptically Filled Products, aseptic process simulations (PDA, 2011):

- evaluates capabilities of aseptic processing operation;
- simulate the aseptic process from the point of sterilization to closure of the container, substituting a microbiological growth medium for the sterile product;
- assess changes made to an aseptic processing operation which might impact the sterility of the final product;
- identify weaknesses in aseptic processing which might contribute to the microbiological contamination of the product;
- evaluate proficiency of aseptic processing personnel;
- comply with current Good Manufacturing Practice requirements; and
- demonstrate appropriateness of operating practices used in support of aseptic processing.

As such aseptic process simulations are a good tool for confirming the aseptic process design. They may also uncover under or unaddressed process variables and weaknesses missed in earlier evaluations. In other words they are a check on the accuracy and effectiveness of assumptions already made. However they are not sensitive enough to rely upon to be the primary judge of the effectiveness of the aseptic process. Therefore *solely* running aseptic process simulations will not:

- *Validate the aseptic process*: The validation of the aseptic process involves a holistic approach, incorporating validation and qualification of numerous aspects of process design and contamination control.
- *Assure process control*: Process control is assured through careful evaluation of the critical quality attributes, process steps, process variables, risks, and control strategies to mitigate those risks.

- *Determine sterility assurance level (SAL)*: SAL, such as those referenced in sterilization processes, cannot be calculated or predicted for aseptic processing by the outcome of media fills, due to lack of statistical correlations between observed parameters and desired outcome. In media fills, each fill is considered as a separate study with a separate set of conditions. Combining those results to fit a statistical model would not be accurate.
- *Qualify clean room personnel*: Participation in successful media fills will not be enough to provide assurance that clean room personnel are properly trained and capable of performing assigned tasks.
- *Qualify support processes and systems*: There are more effective ways to validate support processes, such as wrapping or sterilization of sterilized filler parts, holding conditions for sterile materials, or sterilizing capability of in line filters, than inclusion in media fills.
- *Qualify poor aseptic technique or practices*: Media fills should not be used to attempt to qualify a poorly designed or poorly performed aseptic process. If the media fill passes, that should not mean that such a process step is not acceptable.
- *Train personnel*: Untrained personnel should not participate in aseptic process simulations designed to "qualify" the aseptic process. Trained personnel are a means to eliminate (or reduce) the variability of personnel. In other words, if the aseptic process simulation fails, and the operators were not trained, then those investigating the failure would not know if the failure was due to a process flaw or an operator error.
- *Shake down an unqualified filling operation*: Using an unqualified systems or equipment in a media fill should be avoided for two reasons. (1) If the aseptic process simulation fails, and the line was not qualified, then those investigating the failure would not know if the failure was due to a process flaw or an equipment issue. (2) If the media fill passes, and the line qualification is performed later, then any changes which may occur on the line because of the equipment qualification would not be captured in the simulation and therefore that simulation study may not be valid.

Microbial Contamination Case Study: Sterile Vessel Holding Qualification

To illustrate the importance of understanding the limitation of the aseptic process simulation test capabilities, the following case study example is offered.

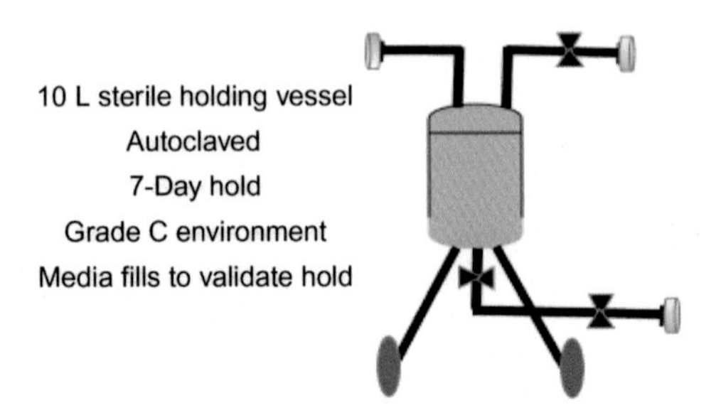

Figure 2.1 Sterile holding vessel.

Background: The company was bringing online a *new aseptic filling operation and clean room suite*. The filling process involved sterilizing product by filtering it from a compounding tank to a sterilized 10 L holding vessel (Fig. 2.1). The vessel was sterilized with moist heat in a qualified autoclave using a validated sterilization cycle. The ends of the vessel line openings were taped and covered with a commonly used, semiabsorbent autoclave wrapping material. Once sterilized the vessel was held up to 7 days in a Grade C environment. The wrapped openings were opened during sterilization and remained open during part of the holding period.

To validate that the interior of the sterilized vessel remained sterile throughout the 7-day holding period, the vessel was held for the maximum holding period and then used as a media holding vessels, during multiple media fills. All media fills passed. No contamination was noted.

Subsequently the facility was qualified and commercial level manufacturing started. Six months later, semiannual media fills were conducted as per the company's compliance policy. Several of those semiannual media fills failed. The source of the contamination was determined to be the interior of the vessel and transfer line.

Question #1: How Did the Interior of the Vessel Become Contaminated?

The answer: The investigation concluded that even though the autoclave had a drying cycle at the end of the sterilization cycle, the vessels were still filled with relatively moist, warm air as they transferred from the autoclave to the cooler Grade C holding area. As they cooled, condensate was produced that collected at the bottom of the vessel. The condensate came in contact with and wets the semiabsorbent wrapping mattering. Once wet the material was no longer an effective microbial barrier. In addition the condensing process resulted in a slight negative pressure in the interior of the vessel. Contamination in the Grade C environment that was drawn to the open end of the vessel contacted the wet wrap material and grew through or permeated to the condensate in the vessel line, thus contaminating the interior of the vessel.

The More Challenging Question Is #2: If the Conclusions Are Correct, Then Why Did the Initial Media Fills Pass?

The answer lies in the assumption of environmental variables. The initial media fills took place during a period when almost no activity was occurring, except for validation. Normal manufacturing had not started yet. The Grade C environment was exhibiting contamination levels closer to Grade A or B environments. During the initial media fills, the Grade C area in which the vessel was held was much cleaner than it would be during normal operations. When normal operations began, the Grade C, normal production activities were occurring in the Grade C area and the Grade C environment exhibited normal levels of activity and contamination. The control system in place to protect the interior of the vessel from contamination turned out to be adequate for the cleaner premanufacturing Grade C environment, but not for the more routine Grade C environment.

Those designing the validation study assumed that the Grade C environment was a constant if it met Grade C specifications. However it turned out that the Grade C environment was a variable even though it continued to meet Grade C specifications. The Grade C environment did not exceed specifications during the semiannual media fills. However, the environmental counts were significantly below limits during the initial media fills.

This case study illustrates two important points. First, that a careful analysis and evaluation of process variables and the risk they may pose on product quality is needed when designing the process. Had that been done, those responsible for the design of the process would have pondered how contamination could appear in the vessel once sterilization is complete. They would have likely concluded that because the contamination could not generate by itself, it would have had to enter the sterile vessel from some point. The obvious weakness to the system would have been the open lines. That would have led to an evaluation for the microbial barriers in place over the open lines and perhaps the realization that the barrier material would not be effective if wet. That might have led to an assessment of the likelihood that the material could have become wet during holding and so on. The conclusion might have been to change the material covering the vessel openings to something less selectable to moisture, to clear the valves sooner, to pressurize the vessel during holding, and to hold the vessel in a cleaner, more controlled area, thus resulting in a more robust control strategy and improved process. Here it would not be the validation study that provides that improvement and increased process performance confidence, but a better understand of process risk and requirements.

The second point is that media fills, while having a definite benefit in providing confidence, that the aseptic practice performance, was not the best tool or method to validate the holding process. Because the process variables they wished to address involved microbial contamination entering the vessel through the wrapped ports, the right tool might have involved an evaluation of the barrier properties of the wrap and the cleanliness of the environment.

Aseptic Process Simulation Study Design

An important question to consider when designing the aseptic process simulation study is whether one is primarily using it to qualify the aseptic filling operator or the process. Process is usually the initial answer given. But decisions made during the design, such as duration of the media fill, inclusion of interventions, and qualification of operators may change depending on what you are trying to do. Human behavior is complex, subject to variability, and unreliable. As such they are difficult if not impossible to validate. The closer one's intent is to qualifying the process, the more effective the aseptic process simulation as a validation tool will be. The closer to qualifying the person, the more problematic and ineffective it will be.

Whatever the design intent is, a written plan subject to quality review should be in place. This plan should present:

- intent and objective of the study;
- clear instructions, written to avoid the potential for individual interpretation, for preference of the study;

- risk-based rationale for study design, configuration, conditions, and parameters;
- objective, quantifiable, and measurable acceptance criteria; and
- provisions for aborting, invalidating, investigating, and addressing deviations related to the study.

A risk assessment should be performed to uncover information needed to make decisions on aspects of study design, such as number of runs, number of units to be filled, including worst-case configurations, fill duration, and intervention inclusion.

Aseptic Process Simulations Performance Schedule and Frequency

There should be written procedures and a risk-based criteria or rationale for the frequency of aseptic process simulation studies. Typically aseptic process simulation studies should be performed for any new or changed aseptic processes. The intent of the study is to confirm that the process or process change achieves its intended outcome and that there are no unintended consequences or residual risks introduced by the process or process change that pose a risk or can adversely affect product quality.

Thereafter aseptic process simulations should be performed periodically so that unintended or missed changes or variables have not been incorporated into the process that poses a risk to product quality. Aseptic processes are complex and it would not be prudent to expect that all changes could be captured all the time. Many health authorities have expressed expectations that these periodic studies be performed on a semiannual basis.

Media Fills Run Number

There should be written procedures and a risk-based criteria or rationale for the number of runs incorporated in the aseptic processing study. Multiple media fills should be run for new and modified or changed processes. Typically three media fills are an acceptable number for multiple run studies. Where changes have occurred to process, multiple runs are prudent to determine the impact of the changes, including unintended consequences of the change and residual risk introduced by the change.

Where there is no evidence of change to the process, as might be the case with periodic aseptic process simulation studies, one run may be sufficient.

There should be enough fills conducted to:

- Adequately challenge the aseptic process and uncover process variables that adversely affect the performance of the process and the sterility of the product.
- Include all process steps, configurations, conditions, and process variables.

Multiple media fills should be performed for new or significantly changed processes. Typically, three media fills are performed. Significant process changes might include:

- changes that increase or result in risk to the aseptic process;
- addition or changes to process steps/interventions;
- complete or near complete change in clean room personnel;
- change in containers or component designs;
- additions or changes to equipment in Grade A or other critical support area or function;
- situations where it is uncertain if significant variables have been introduced to the process—including after prolonged shutdowns, where extensive maintenance has been performed, extensive maintenance performed other than during shutdowns, or where adjacent areas have undergone changes or extensive maintenance; and
- other indications of process variance or failure.

Multiple media fills may also be recommended where Corrective Actions and Preventive Actions (CAPAs) have been employed for previous process or aseptic process simulation failures. In this case, the media fills are run to confirm that the CAPA has been effective in addressing the issues and that there are no unintended consequences or residual risks added to the process because of the changes or CAPA.

A risk assessment can be used to help decide the number of media fills warranted. If one is uncertain whether the impact of the change is significant, then one should default to multiple media fills. Even if one thinks that a change has little or no impact, it is difficult to be certain without confirming through assessments that may include multiple media fills.

If there is little or no evidence of change or effect on the process by the change or actions, then a single media fill may be warranted. Situations where a single media fill is appropriate may include:

- where media fill failure is not attributed to the aseptic process;
- failed growth promotion studies;
- aborted or invalidated media fills;
- large-scale container closure failures;
- at the end of prolonged process or clean room shutdowns where extensive maintenance or changes have not be performed; and
- periodic repeat of media fills.

Once initial aseptic process simulation studies have been completed, it is recommended that media fills be repeated on a periodic basis. Many, if not most, global health authorities recommend or expect semiannual (every 6 months) repeat of media fills. Typically these are single media fills for each aseptic process configuration. Periodic media fills have scientific value because it is difficult to predict or know of all the changes or variables that might arise in a complex process such as aseptic processes. Periodic media fills can be run as single runs because there is no evidence of process changes.

Inclusion of Process Steps

An aseptic process simulation should be designed such as any process step that may pose a risk or adversely affect the sterility of the product should be included in the aseptic process simulation, unless those steps are adequately validated by some other means or study.

Aseptic process simulations should include such process steps as:

- aseptic additions during compounding;
- product transfer down stream of final sterilizing method (or filter);
- material and component transfer into the critical area;
- fill or aseptic manufacturing line setup;
- aseptic connections—including those to sterilizing filter outlets;
- special assemblies—such as PUPSIT (poststerilization, preuse integrity test) assemblies;
- inherent (routine) and representative corrective (nonroutine) interventions;
- filling and sealing operations;

- environmental monitoring in or adjacent to critical areas;
- fill checks conducted in the aseptic processing area;
- shift changes, where the movement of clean room personnel might affect or disrupt environmental conditions or process performance;
- component loading;
- intervention response sanitization; and
- lyophilized product transfer and cycle mechanical operations.

"Worst-Case" Parameters or Conditions

The rationale for the selection of worst-case configurations, parameters and conditions for the ascetic process summation should be written into the protocol or some other procedure or study document, subject to quality review. Worst-case in the context of aseptic process simulations does not necessarily mean the worst possible case. It means the process parameters and conditions that are most likely to uncover weakness in the process. To obtain the best scientific value from the study, worst-case conditions should be used in the process simulation studies. It can be surmised that if these conditions are challenged in the media fills, then conditions representing less than worst-case should be qualified as well.

In theory, this can be an effective way to design aseptic process simulations. However it is not always easy to determine the actual worst-case parameters and conditions. For example, would the worst-case be the maximum or minimum number of operators in the clean room? Maximum number of operators represents a larger source of microbial contamination, yet minimum number may mean that operator movement is increased, and focus decreased. Both may represent worst case. In another example, consider which might be worst-case largest container or smallest container. The largest container would have the larger opening, affording more opportunity for contamination to enter, yet the small container may be less stable during the transport and filling process. Again it may not be clear which represents worst case. A final example would be slowest versus fastest fill speed. Slower fills represent more opportunity for contamination prior to sealing; however, faster fills may represent more mechanical movement, potential airflow disruption, human activity, and unstable container transport. If one cannot choose the worst case from two or more sets of parameters and conditions, then both or all the potential worst-case configurations should be included in the process simulations study.

Fill Volume

The minimum quantity of media filled into the container should be sufficient to contact all interior surfaces and enough to show indications for contamination upon visual inspection. If possible, it is recommended that the normal production volume be filled to address any risk posed by full fills, such as foaming and splash up onto fill needles. Full fill volumes or fill volumes that approach full container volume should be qualified through growth promotion studies to confirm that sufficient oxygen is present to support low levels of microbiological growth.

Duration

Media fills should be long enough to capture all activities, interventions, and actions relevant to the maintenance of aseptic process performance and product sterility. Media fills should be long enough to fill enough units to capture the low levels of contamination that may be present or inherent in the aseptic process. Media fills should be long enough to properly challenge the process to discover any flaws, weakness, or underaddressed variables into the process that can pose a risk to process performance and product sterility. The duration of the media fill will, in part, depend on the length of commercial production runs. The duration of the media fill should be based on scientific, risk-based, and logical evaluation of the objective of the study. The duration of the media fill should not be used to prove the acceptability of process conditions, where simulations are an ineffective means to do so, or where those process steps represent poor aseptic practice.

Full duration (or longer) media fills may be of value. However risk evaluation may show that full or longer duration media fills are not necessary, where no valuable scientific information is gained from running full duration. In some cases, longer duration media fills may lead to a false sense of security, where more effective ways to qualify process conditions are ignored in favor of merely running the long duration fill.

Having said that, longer duration media fills may be an effective study design element, where duration-related activities and conditions can only be included and evaluated during a longer run. Risk assessments are an effective method for acquiring information that will be helpful in making informed scientifically valid decisions on the duration of the media fill.

In the end, it is important that the design of the media fill take into consideration the objective of the study and whether and how the duration of the media fill will help meet that objective.

For example, what is the value of running a fill duration media fill? One might answer that full duration is required to determine the maximum length of time the production fill can run. But how does the duration of the media fill meet that objective? While it is a true simulation at full duration, how does that meet the objective of proving confidence that the process will be reliable?

For instance—two variables that one may try to address with a long duration fill are: (1) whether the environment remains suitable for aseptic processing during the entire length of the fill and (2) whether human fatigue becomes an issue. Let us explore each.

Clean room environmental deterioration: The concern is that the clean room environment may deteriorate over the course of the fill to the point where it is no longer suitable for the aseptic process. First, is the media fill the best way to determine or uncover this condition? Wouldn't the design of the clean room, process, and heating ventilation and air conditioning (HVAC) system, along with environmental monitoring program, including temperature, humidity, differential pressure monitoring a viable and total particulate sampling, be a more effective means to do so? Second, is the media fill sensitive enough to determine if the clean room environment has changed to the point where it will pose and unacceptable risk to product quality? It probably is not an effective means to do so. The media fill is not designed to monitor or judge the condition of the environment.

Finally, does the environmental monitoring program indicate that a properly designed and operated clean room environment worsens over time? It probably does not. And if it did, would media fill passage be the proper response to confidence that a worsening clean room environment was acceptable? or would the company be better off taking appropriate actions to correct the worsening environmental conditions?

The effect of human fatigue: Properly designed and operated clean rooms may not get worse over time, but it is undeniable that people do get tired. The problem with using media fills to determine the effect of human fatigue on the aseptic process is that it is impossible (or very difficult) to simulate human fatigue in the clean room operation. Allowing operators to perform in media fills for extended periods of time does not simulate all the variables associated with fatigue. For instance, it does not consider how much rest or sleep the operator had the night before, or any nonwork-related stress they might be under, or levels of extraordinary efforts they may be making because they are aware that they are participating in a highly observed test.

The best way to address the process variable of human fatigue is by reducing the risk or potential effect human fatigue may have on the outcome of the process. In other words, by eliminating the cause(s) of fatigue through ergonomics, better process design, increased breaks, and such. It is not by running and passing a media fill.

For example, allowing an operator to lift heavy trays of bottles for six hours in the clean room is tiring and there is a risk that as the day goes on and they become more tired, those trays will be carried closer and closer to their bodies, thus increasing the risk of microbial contamination. The right approach would be to allow for more frequent breaks to enable the operators to replenish needed strength and/or to lessen the weight of the trays or empty a more automatic system for transporting the glassware. Compare this response to running three full duration media fills, where operators are performing these knowingly strenuous activities, that pass. Which approach would give you more confidence in the robustness of the process. Again, are we using media fills to qualify the person or the process? Are we using them here to qualify the "strength and endurance" of the operator? If so, then is strengthen and endurance criteria for clean room operations? One can see how problematic this might be.

One final point, if taking rest breaks is found to be a way to mitigate the risk of operator fatigue, and if operator fatigue is thought to be an important process variable, then the inclusion of breaks should be part of the media fill design.

Therefore the *minimum* duration of the media fill should overlap the time the operator is in the clean room between break periods and should include the transfer of operators during those break times.

It should also be noted that many global health authorities and regulatory agencies, including the US FDA and EMA, have minimum recommendations and expectations for the number of units filled based on the quantity of commercial batches anticipated (FDA, 2004).

Interventions

Interventions, usually, are human manipulations or activities that occur during the aseptic process in the proximity to sterile or product contact surfaces and as such may pose a risk to product quality and aseptic process performance. As such they may pose a potential risk to product sterility.

One should not rely merely on media fills to qualify or prove that interventions are proper or pose no risk to exposed sterile product or surfaces. The qualification of interventions begins with proper design of the activity in regard to protecting product from this risk of contamination. Interventions should be designed to adhere to proper contamination control and first air principles. Interventions should be performed by trained personnel, properly gowned, using sterilized materials, tools, and utensils (where applicable), with at risk containers, surfaces, and product removed or decontaminated subsequent to the intervention.

Interventions should be included in the design and performance of the media fill process simulation study, as they are an important part of the aseptic process.

There are two general categories of interventions. Inherent, sometimes referred to as routine, interventions are those interventions which are an integral part of the aseptic process. They are inherent in the process and the process cannot be performed without these interventions happening. These may include fill line setup, fill weight, or volume checks, adding of components, environmental monitoring.

The second category of interventions is corrective or sometimes referred to as nonroutine intervention. These are interventions that are proper to include in the aseptic process but do not necessarily have to occur during the process. These might include such interventions as removal of a fallen vial, dislodging of jammed stoppers, clearing of a broken vial, and so on. All interventions should be conducted using proper aseptic technique. Validation studies, such as media fills, should not be used to try to qualify an intervention that relies on suspect technique.

A risk-based approach should be used to assess interventions and decide how interventions will be included in the aseptic process simulations and media fills. Risk assessments can help those involved with designing the study decide on the following:

- Relative risk of an intervention.
- Steps required to reduce or address risk posed by the intervention.
- Which interventions are included in the media fill or fills.
- How often the intervention will be performed during the media fill or fills.
- Which operators will be required to perform interventions during the media fill or fills.
- The procedure for addressing an intervention that has not been included in an aseptic process simulation.

One should also consult regulatory (FDA, 2004) and industry (PDA, 2011) guidance when designing aseptic process simulation studies.

Intervention Evaluation and Risk Assessment Methods

Companies often use intervention ranking methods to categorize or group interventions to decide which interventions are included in media fills. One such method is the I-REM (Baseman et al., 2016a,b,c) or Intervention Risk Evaluation Method. I-REM is an objective model for evaluating the relative risk of aseptic processing interventions (Moldenhauer and Madsen, 2014).

Aseptic process risk assessments are sometimes difficult to perform, because of:

- The high level of severity or impact to product quality associated with process failure.

- The lack of clear correlation between what can be observed and the desired or undesired outcome.
- The relative difficulty and unreliability of process failure detection.
- The complexity and interdependence of process steps and conditions include human behavior and performance.

These issues can result in a lack of objectivity that make it more difficult to perform effective aseptic processing risk assessments. Therefore aseptic process risk assessments have tended to be more subjective and of limited benefit. An objective method for evaluating risk of aseptic processes would allow for better process understanding and aseptic process simulation-related decision making and design.

The I-REM is an objective risk-based method that identifies the intervention-related variables and factors that could result in process failure and the loss of sterility. The risk question here is what is the likelihood and impact of such a failure? To what extent does the performance of an intervention or the failure of an intervention to be performed as planned, perhaps by human error or by process flow, pose a risk to the sterility of the product? The I-REM is based on answering three key questions:

1. What are the factors that define the risk of the intervention?
2. What are the criteria by which these factors can be measured?
3. How should these multiple factors be considered to define overall intervention risk?

The I-REM attempts to avoid subjective values and instead emphasize quantifiable criteria for determining risk values. As such, it is designed to be:

- *Objective*. Objectivity reduces the potential for bias. For the I-REM, objectivity is defined as the extent that different people with similar knowledge about the process and access to information are able to use the I-REM and obtain a similar risk rating for a given intervention. For this to happen, meaningful, measurable data, or information must be available.
- *Simple*. Simplicity allows for the use of the I-REM by all interested members of the aseptic processing team. This leads to better acceptance of the I-REM outcome. For this to happen, the data or information must be easily accessible.

- *Robust and reproducible.* The method should be equally applicable to all interventions in all aseptic processes, whether they are performed in manual, conventional, restricted access barrier systems, or closed isolators' aseptic filling processes.
- *Logical and defendable.* The I-REM should be based on sound science to be viewed as truly objective, to gain user acceptance of results, and convince regulators that the method and its results are credible.

The I-REM uses a *Keyword approach*, with the people assessing the process choose the criteria and words that have the most meaning to them in regard to meeting the line of sight objective. The Keyword approach allows one to better utilize the individual company team's knowledge to the problem at hand by selecting the words and terms that are most meaningful and clear to them. This may be accomplished by using a brainstorming session with I-REM risk assessment team members.

For example, the I-REM risk assessment team selected the common intervention risk factors. In this example, they determined measurable criteria for each factor:

Duration, defied as the amount of time it takes to perform those steps of the intervention that could pose a risk to product sterility—as determined through review of media fill and manufacturing observation logs.
Complexity, defined as the complexity or difficulty in performing those steps of the intervention that could pose a risk to product sterility—as determined from SOPs.
Proximity, defined as the proximity of the person performing the intervention to exposed sterile product or product contact surfaces—as determined from observation.

It is important to note that no one risk factor defines relative risk; it is the combination of all risk factors. Different analysis tools are used depending on whether all risk factors are weighted the same or if one or more have greater impact on product quality or process performance.

In the example shown in Table 2.1, the ranking criteria was set as Medium or normal being where most (approximately 80%) of all interventions fall. Any measurement above normal was considered of higher concern and therefore HIGH ranking and any measurement below normal considered to be of lower concern or LOW ranking.

Once the criteria are set, individual interventions can be ranked using such tools as a two-table triage system, as shown in Tables 2.1 and 2.2. This method ranks interventions into three categories: Low-, medium-, and high-risk rankings (Table 2.2). The method can be expanded to include more categories if needed. The importance of an objective ranking method for ranking interventions is that it provides a risk-based criteria that may be used to decide on the inclusion of interventions in media fills, the frequency of the inclusion, and whether all operators will perform each intervention during the media fill.

Incubation

Filled media units should be inspected prior to incubation. Units found to have defects that compromise the integrity of the sealed container,

Table 2.1 Risk Factor Ranking Criteria.

Risk Factor Level	Duration	Complexity	Proximity
High	More than 10 min	More than five steps	Operator breaks first air with nonsterile entity
Medium (normal)	Between 1 and 10 min	Between two and five steps	Operator breaks first air with sterile entity
Low	Less than 1 min	One step	Operator does not break first air

Table 2.2 Two-Stage Risk Assessment Tables.

		Duration		
Complexity		Low	Medium	High
	High	Risk Class 2	Risk Class 1	Risk Class 1
	Medium	Risk Class 3	Risk Class 2	Risk Class 1
	Low	Risk Class 3	Risk Class 3	Risk Class 2
		Proximity		
Risk class		Low	Medium	High
	Risk Class 1	Medium risk	High risk	High risk
	Risk Class 2	Low risk	Medium risk	High risk
	Risk Class 3	Low risk	Low risk	Medium risk

For a more complete discussion of the I-REM method, refer to the references provided.

such as holes, cracks, gaps incomplete seals, dislodged, or missing stoppers should be removed prior to incubation. Incubation of such nonintegral units would not yield scientifically valuable information, because contaminated units would likely be blamed on the nonintegral nature of the container. Nonintegral units that are not uncovered and rejected and are incubated and later found to be contaminated should be considered contaminated units and must be investigated and addressed as such.

Incubation temperatures between 20°C and 35°C for 14 days are often recommended by health authorities and are frequently used. The 14-day incubation period is designed to allow for slow growing microorganisms. Media filled units may be removed from incubation to be checked at intervals during the 14 days. Where this is the case, time out of the incubator should be added to the total incubation time.

The PDA Points to Consider recommends one temperature between 20°C and 35°C, because media fills are qualitative analyses and most mesophilic microorganisms will grow at that temperature range. However, companies choose to use more than one temperature for incubation. For instance, multiple temperatures of 20°C–25°C and 30°C–35°C for 7 days each are sometimes used. Where media filled, units are incubated sequentially at multiple temperatures; the sequence of temperature incubation (high then low, or low then high) should be justified, specified, and recorded. Companies should choose incubation conditions and temperatures that provide the best opportunity to uncover anticipated sources of contamination. Understanding anticipated sources of contamination may be based on such factors as potential sources inherent in the process design, for example, human exposure, packaging materials, moisture or from environmental isolates (FDA, 2004; PDA, 2011, 2015, 2016).

Growth Promotion Studies

The ability of the media to support growth of relatively low levels of a wide range of microbiological contamination should be demonstrated through growth promotion studies. Studies should be conducted initially to qualify the type and supply of media and the incubation conditions—time, temperature, and sequence if more than one temperature is used.

Thereafter each batch of media used in a media fill should be qualified by selecting samples of filled media units and inoculating low levels of contaminates (<100 cu) selected from environmental isolates and/or indicator organisms: *Candida allicin* (ATCC No. 10231), *Aspergillus brasiliensis* (ATCC No. 16404) (formerly *Aspergillus niger*), *Escherichia coli* (ATCC No. 8739), *Pseudomonas aeruginosa* (ATCC No. 9027), *Staphylococcus aureus* (ATCC No. 6538) (United States Pharmacopeia, 2014).

These ongoing growth promotion tests should be performed on filled media containers selected from the end of their incubation period. In this way, the media has gone through the same conditions as all the media in the aseptic process simulation study. This also eliminates the potential of selecting a contaminated unit prior to incubation, which could compromise the validity of the test and study.

Preincubation Inspection and Rejection

Filled media units that have cosmetic defects but are otherwise integral should be incubated.

The criteria for rejection should be specified in written procedure or instructions and the reason for rejection should be noted in the protocol or batch record. Inspectors should be trained in the inspection of media filled units.

Containers that are normally removed as a result of interventions may be incubated if their evaluation would yield useful scientific information about the intervention or technique used during the intervention. For example, if the intervention is conducted using proper aseptic technique and no contamination of adjacent containers is expected—then the incubation may yield valuable information, if contamination does occur. However if the intervention is more intrusive and higher risk then incubation of normally removed containers would be expected. In this case, incubation may not yield valuable information, as the cause would be assigned to the intervention. It may also be difficult to denote, and segregate containers automatically rejected as part of the filling operation. If filled media units, that are normally rejected during commercial filling, are rejected, then it should be demonstrated that the procedure or mechanism for removal of such units is robust and reliable.

A well-designed risk assessment may be used to help decide the criteria by which filled media units are included or excluded from incubation.

Postincubation Inspection

The filled media units are inspected at the end of the prescribed incubation period. Media units may be inspected on an interim basis prior to the end the full incubation period, providing the units are then returned to the incubator for the full incubation period. Where media units are removed personnel performing the inspection should be trained, qualified, and periodically checked to uncover relatively low levels of contamination. These inspectors should be under the supervision of the microbiologist or microbiology laboratory during the media inspection procedure. Microbiological contamination may appear as turbidity, globules, or filaments. Any such evidence of contamination should be noted as to tray and time found. These units should be cultured as soon as possible to confirm and identify the source of microbiological contamination. This is also important because what initially appears as microbiological contamination may be found to be nonmicrobiological contamination, such as chemical or other foreign material.

Acceptance Criteria

Because the objective of aseptic processing is to manufacture sterile products with no microbiological contamination, the acceptance criteria for the media fill should be zero positives. While the presence of one contaminated unit may not necessarily prove that the process is incapable of manufacturing sterile product, it is an unwanted event that should be thoroughly investigated and evaluated.

In addition, because the aseptic process simulation is performed after nearly all other system and process qualification runs are complete, it is important that the process is ready for commercial manufacturing. Therefore the line should run well. Issues with line operation, elevated reject and defect rates, and poor yields are indicative of the line not being ready.

Filled Unit Accountability

Media filled units prior to incubation may be reconciled to the same limits as commercial production. However, once incubated, media filled units should be 100% reconciled. The distinction is that it can be argued that there may be a temptation to discard or lose a contaminated unit. There would be no such temptation prior to incubation, because the units would not indicate contamination. However after incubation, they might. The issue is that if it is necessary to have a 100% reconciliation limit on media filled units leaving the incubator, then one must know the exact number of units entering the incubator. Therefore it would be prudent to have a 100% reconciliation on all media filled units regardless of whether they are incubated or not.

Failure Investigation

All contaminated or positive media filled units must be investigated to determine the source of the contamination. Once a failure is noted, it is not proper to continue aseptic process simulation studies—until three runs pass. If may not be possible to find the definitive root cause of the contamination. However an assignable or plausible cause for the contamination should be determined.

The following are some helpful investigation hints one may want to consider:

- Culture and identify microorganism(s) contaminating positive media filled units.
- Follow the microbiological clues—common environmental monitoring isolates, potential or likely sources of contamination (i.e., human borne, water borne, corrugated borne, etc.).
- For ongoing aseptic process simulations, where media fills have passed in the past, try to determine what may have changed, what is different, what new variables have been introduced into the process.
- Complete the full investigation. Avoid stopping after one or first cause is found.
- List all possible causes, then attempt to eliminate each cause based on evidence.
- Any possible causes that cannot be eliminated should be treated as a plausible cause.
- Correct each plausible cause with a CAPA.

- Address the impact of failures on product filled on the effected line/process and similar lines/processes.
- Follow-up on CAPA to ensure that CAPA actions are completed in a timely and effective manner.
- Modify risk assessments to take into consideration information learned from failure investigations.

Aborted and Invalid Media Fills

Aborted media fills are those that are stopped once they have begun. Invalidated media fills are those that are completed but results not used to evaluate the process.

Media fills may be aborted if conditions arise during the fill that would result in commercial production fills to be stopped. These conditions and the criteria for stopping production and media fills should be similar and should be captured in a written procedure. Examples might include fill line or HEPA system break down; interventions that would normally stop the fill, discovery of information that would compromise the ability of the media fill to yield valuable scientific information and was not known at the start of the run.

Media fills may be invalidated if conditions or information becomes known that would result in no valuable scientific information from the media fill. Examples might include failure of media to support growth, failure to accurately simulate the process, residuals, or carryover of materials that could compromise the ability of the media to support growth.

Because aborted or invalidated media fills are undesirable and may reflect production issues, they may require investigation to determine and reduce the likelihood of future causes.

Special Considerations

This chapter has focused primarily on aseptic processing liquid fills. Parenteral products also include nonliquid fills, including powder and ointment filling, as well as lyophilized products. Although many of the principles and methods presented in this chapter are applicable to those other types of manufacturing, there are special considerations for those processes. These considerations will involve good understanding of the process and risks, and well thought out, often quite unique approaches to filling, procedures, and conditions to best simulate the process.

Powder Filling

Powder filling processes present a challenge in that the filling of liquid media does not simulate the process. The equipment and systems in place are designed for dry powder filling, not for liquid media filling. Therefore special means must be employed to expose the process to liquid media. One option is to fill liquid media through the powder filling equipment. However dry powder fillers may not be designed to fill liquid. Another option is to add a liquid filler to the fill line, usually after the powder filler. In this way, the containers can be filled with a nonbacteriostatic powder simulating the actual filling operation and then liquid media can be added for the challenge. However this involves adding a separate piece of equipment not normally on the filling line or involved with the aseptic process. This adds risk to the media fill, not inherent in the commercial operation. A third option is to fill the containers with a nonbacteriostatic powder and add sterile media after the filling on a unit by unit basis. However this method adds considerable risk of contamination. All methods for aseptic process simulations for sterile powders involve some risks and disadvantages. It is important to understand those risks to design and evaluate the simulation studies that take the challenge of these risks into consideration.

Ointment Filling

Some of the same issues of filling liquid media inherent in powder filling are present with ointment filling. In addition, ointment filling is often performed using product at elevated temperature to allow for the ointment to flow more consistently. The elevated temperature could greatly reduce the capability of the media to support growth. Therefore if a growth promoting media is used, it should not be filled at that elevated temperature.

Lyophilized Product Filling

Lyophilized products are usually filled in the same manner as nonlyophilized liquid products with the exception of stopper placement and lowering. Lyophilized product filing should take into consideration:

- The transport of semiopen vials from the filling line to the lyophilizer.

- The loading of the vials in the lyophilizer.
- The operation of the lyophilizer, including condensers, compressors, and shelf movements.
- The effect of the lyophilizer cycle on the capability of the media to support growth, including the freezing cycle and the use of nitrogen overlay. It is recommended not to freeze the media, but to simulate the freezing process. It is recommended to substitute sterile compressed air for the nitrogen.

Anaerobic Processes

Most microorganisms present in the clean room thrive in aerobic conditions, where ample levels of oxygen are present. However there are some microorganisms that thrive in anaerobic or oxygen depleted environments. Where the aseptic process is anaerobic, special media, such as fluid thioglycolate may be required to allow for growth of those microorganisms.

These anaerobic processes are relatively rare in open filling suites. There may be such processes involved with closed systems aseptic recrystallization processes and other sterile API manufacturing processes. Generally nitrogen flush and sparing procedures do not deplete or replace enough oxygen to create anaerobic condition, so the use of anaerobic media would not be needed. However where nitrogen flush and sparging are used in the filling, holding, or transfer procedure, that nitrogen should be replaced with sterile compressed air.

Conclusion

To be effective, the design of aseptic process validation must use a risk-based holistic approach. Many of the procedures and process steps can be qualified or validated using standard validation methods, including those for leaning, sanitization, equipment function, sterilization, filtration, container closure integrity, inspection, labeling, and transport. The qualification or validation of the unique element of the aseptic process, the aseptic practices and technique is best validated by careful design of those procedures and processes. The confidence of the reliability of performance begins and is best demonstrated by well thought out, risk-based design, and planning. One cannot test quality into a product and one cannot validate performance into a process.

Designing a successful validation approach to meet the process control objectives should be based on five key principles:

1. a good understanding of the relationship between product quality, process risks, and control strategies;
2. recognition of the effectiveness and limitation of the validation study;
3. choosing the right study and tests to meet the process control objective;
4. relying on process design as the primary tool to establish process confidence; and
5. learning from the outcome of studies in order to improve the process.

The validation aseptic processes involve the identification of process function, variables, and control strategies designed to ensure success of the process. Aseptic process simulations, encompassing media fills are one tool for evaluating the processing steps used to manufacture a sterile product. Since an aseptic process incorporates many systems to assure and control sterility of the materials produced, a holistic approach is used to control aseptic processes.

The approach to validating the aseptic process may include: product, equipment, and component sterilization; personnel training and certification; equipment and facility sanitization programs, equipment and facility operation; environmental controls; personnel, material, and equipment flows; and overall process design. Each control strategy and system that affects the quality of the process and product should be evaluated and tested during the validation approach.

Aseptic process simulations cannot validate the aseptic process alone. However, they are an important tool to be used. The aseptic process simulation allows one the opportunity to uncover variables missed in or not taken into consideration during the design of the process. While, not the only or best way to ensure process capability, when designed and planned properly, they can be a very strong indicator of both process control and weakness.

References

Baseman, H.S., Stabler, M., Henkels, W., Long, M., 2016a. A Line of Sight Approach for Assessing Aseptic Processing Risk: Part I, PDA Letter. May 31.

Baseman, H.S., Stabler, M., Henkels, W., Long, M., 2016b. A Line of Sight Approach for Assessing Aseptic Processing Risk: Part II, PDA Letter. July 7.

Baseman, H.S., Stabler, M., Henkels, W., Long, M., 2016c. A Line of Sight Approach for Assessing Aseptic Processing Risk: Part III, PDA Letter. September 27.

FDA, 2004Guidance for Industry, Sterile Drug Products Produced by Aseptic Processing—Current Good Manufacturing Practice, U.S. Department of Health and Human Services Food and Drug Administration, Center for Drug Evaluation and Research (CDER), Center for Biologics Evaluation and Research (CBER), Office of Regulatory Affairs (ORA). September, Pharmaceutical CGMPs.

Moldenhauer, J., Madsen, R. (Eds.), 2014. Contamination Control in Healthcare Product Manufacturing, Volume 3. PDA, Bethesda.

PDA, 2011. Technical Report No. 22 (Revised 2011) Process Simulation for Aseptically Filled Products, PDA, Bethesda.

United States Pharmacopeia 2014. Chapter USP <51>, ORA Pharmaceutical Microbiology Manual.

PDA, 2015. Points to Consider for Aseptic Processing, Part 1.

PDA, 2016. Points to Consider for Aseptic Processing, Part 2.

CHAPTER 3

Quality Risk Management of Parenteral Process Validation, Part 1: Fundamentals

Amanda McFarland
Valsource, Inc., Downingtown, PA, United States

Quality Risk Management and Process Validation

Overview: Risk Management Evolution

In 2006, *ICH Q9, Quality Risk Management* (ICH Q9, 2006) introduced the concepts of risk management to the pharmaceutical industry and is the foundational document available to our industry on risk management concepts. The concepts surrounding risk management (i.e., risk identification, risk control and prevention, and risk review) have long been utilized by industries for informed decision making and failure avoidance including economic, automotive, and aeronautical. With the introduction of these concepts to the pharmaceutical field, many benefits presented themselves including the opportunities to demonstrate to the health authorities robust process understanding, roadmaps for focusing on the most vulnerable portions of manufacturing processes and a concrete mechanism for knowledge management and knowledge transfer. ICH Q9, however, is not the only guiding document available to our industry to benchmark risk management ideas and concepts. *ISO 31000 Risk management—Principles and guidelines* (AS/NZS ISO, 2004) outlines the critical areas to consider using the risk management process and provides guidance regarding risk management concepts which can be applied against organizations of any type or size. Before ICH Q9 and ISO 31000 were published and adopted by the pharmaceutical businesses, the medical device industry was executing risk management and implementing programs as part of their compliance with *ISO 14971 Medical devices—Application of risk management to medical devices* (ISO 14971, 2009), which was first published in 2000. For this reason, medical device organizations have matured their risk management programs overtime and have adapted risk management tools and employed risk-based thinking to suit their industry specific needs.

Principles of Parenteral Solution Validation. DOI: https://doi.org/10.1016/B978-0-12-809412-9.00002-2

In 2009, the full intent of ICH Q9 was revealed through ICH Q10 *Pharmaceutical Quality System* and outlines the role of QRM and knowledge management as enablers of the Pharmaceutical Quality System (PQS). This function highlights both the importance of integrating risk management into each quality system and demonstrates how the integration creates an environment of knowledge sharing and preservation of scientific knowledge overtime.

In 2011, the FDA published their perspective on QRM as it relates to Process Validation. In *Process Validation: General Principles and Practices* (Guidance for Industry, 2011), validation experts were provided with the FDA's perspective on the use of risk management to facilitate Process Validation activities. This guidance document leveraged the insight from *ICH Q8 (R2) Pharmaceutical Development* (ICH Q8, 2009), *ICH Q9 Quality Risk Management* (ICH Q9, 2006), and *ICH Q10 Pharmaceutical Quality System* (ICH Q10, 2000) to support a life cycle approach to drug development and the process validation activities integral during a product's life cycle. Within this guidance document, Process Validation is defined as "the collection and evaluation of data, from the process design stage through commercial production, which establishes scientific evidence that a process is capable of consistently delivering quality product." (Guidance for Industry, 2011). To this end, the ability to consistently deliver quality product, is continuously referred to within the guidance document as a foundational principle of developing validation schemes and is directly supported with the use of risk-based thinking and QRM.

The FDA's guidance document defines three phases of process validation. These stages are illustrated in Fig. 3.1.

Quality and patient safety are evaluated within Stage 1. In this phase, critical quality attributes (CQAs) are defined and serve as the foundation of the subsequent stages of validation. CQAs are the data points measured throughout manufacturing that give us confidence in the quality of the process and our ability to protect the safety of the patient. As process understanding and the CQAs are defined, process design and process understanding are further explored to determine what critical process parameters (CPPs) are required to ensure that the CQAs will be met. It is important that CQAs are both well documented and defined using a risk-based strategy (i.e., risk assessment or through criticality analysis) with a clear link to the parameters in place

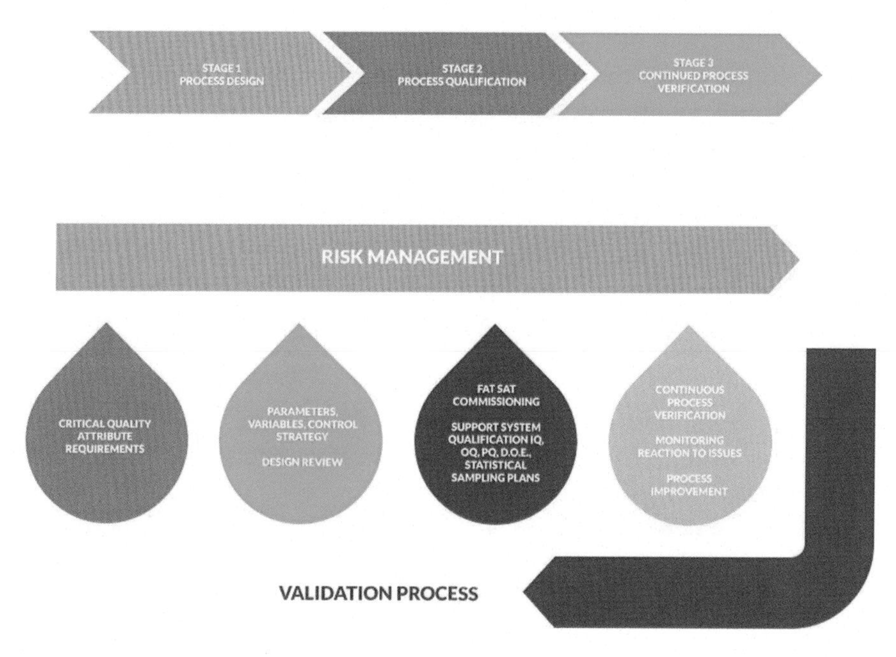

Figure 3.1 Stages of process validation.

to ensure the CQAs remain in a state of control. Stage 1 also serves as the opportunity to understand the variability in the process and design control strategies that limit the amount of variation throughout the process.

Understanding and then controlling the variables through a risk-based control strategy leads to Stage 2, Process Qualification. During process qualification, a risk-based sampling strategy will ensure that all parameters critical to process success are appropriately monitored and tested. A well-designed process and the knowledge of that process will yield the confidence to achieve the ultimate goal of providing a safe, effective and consistent product to the patient.

Stage 3 offers an additional of layer of assurance during the commercial manufacture of the product. This stage requires that the data obtained during commercial runs are continuously evaluated to determine if additional factors contribute to variation and if so, what additional process controls are to be invoked implemented to control that variation. The intent of continuously monitoring the process and adjusting the process to account for variation is the opportunity for continuous process improvement.

The life cycle of process validation, much like the QRM life cycle, requires that the knowledge obtained overtime be not only be reviewed and monitored but that action is taken to account for any variation identified overtime. The underlining mechanisms driving process validation are the principles of risk management and risk activities which serve as checks and balances to implement change and ensure a quality product is produced using a robust and reliable manufacturing process. This chapter will describe the Quality Risk Management Life Cycle and provides strategies for completing successful risk activities.

Principles of Risk Management and the Quality Risk Management Life Cycle

Risk Management is not a singular activity; it is the combination of activities in a life cycle approach which gives it value and from where knowledge is derived. An additional benefit is exploring uncertainty and the efforts to minimize uncertainty present in the manufacturing process. Depending upon the source, various definitions of risk management are available to consider when developing a risk management program. ISO 14971 defines risk management as the "systematic application of management policies, procedures, and practices to the tasks of analyzing, evaluating, controlling and monitoring risk" (ISO 14971, 2009). The risk management process in ISO 31000 provides a similar definition with additional support of the communication and consulting elements of risk "systematic application of management policies, procedures, and practices to the activities of communicating, consulting, establishing the context, and identifying, analyzing, evaluation, treating, monitoring, and reviewing risks." ICH Q9 and ICH Q10 define QRM as "a systematic process for the assessment, control, communication, and review of risks to the quality of the drug (medicinal) product across the product life cycle" (ICH Q9, 2006; ICH Q10, 2000). As noted in the definitions above, multiple activities are executed within the risk management process. The core elements include Assessment, Control, Communication, and Review, which are demonstrated as a life cycle in Fig. 3.2. Fig. 3.2 provides demonstrates an example of the life cycle nature of QRM.

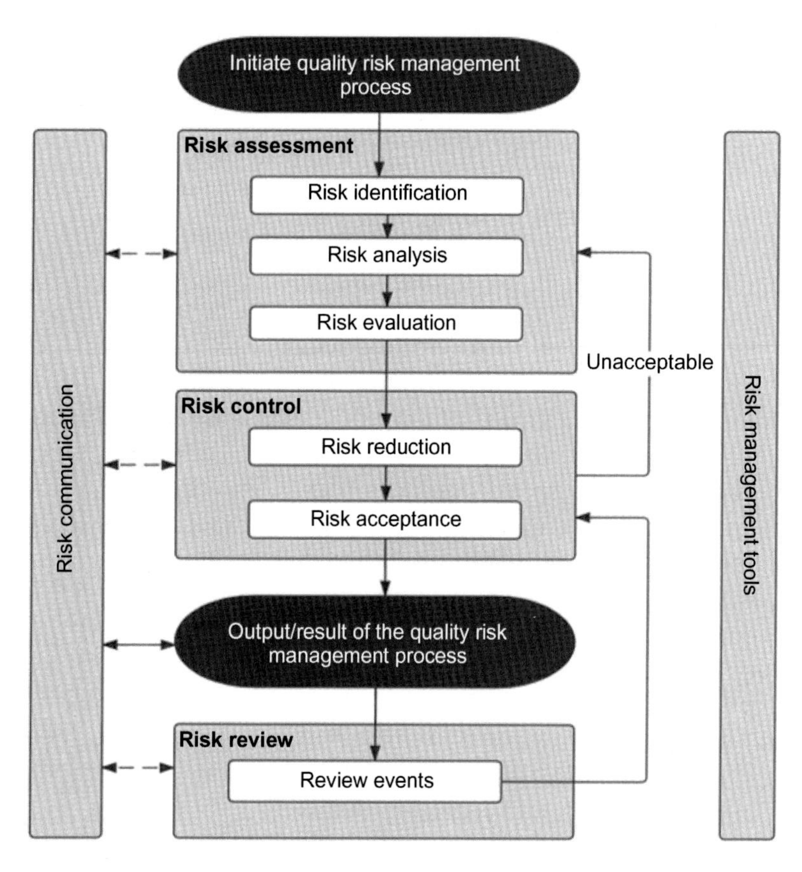

Figure 3.2 Quality risk management life cycle.

Risk: Understanding Hazard, Harm, and Controls

One of the most valuable lessons when furthering knowledge of risk management is understanding what "risk" really means. Below are definitions of risk that are common to the pharmaceutical industry as well as a definition common to our everyday experience with risk.

Source	Definition of Risk
ISO 31000	Effect of uncertainty on objectives
IEC Guide 51	Combination of the probability of occurrence of harm and the severity of that harm
ISO 14971	
ICH Q9	
ICH Q10	

(Continued)

(Continued)

Source	Definition of Risk
Merriam-Webster	1. Possibility of loss or injury 2. Someone or something that creates or suggests a hazard 3. (a) The chance of loss or the perils to the subject matter of an insurance contract; *also*: the degree of probability of such loss. (b) A person or thing that is a specified hazard to an insurer. (c) An insurance hazard from a specified cause 4. The chance that an investment (such as a stock or commodity) will lose value

Each of these definitions has one thing in common: preserving *things of value*. *Things of value* or key elements to explore are defined in the risk question and reinforced with the development of risk ranking criteria. However, identification of "risks" is more than just indicating that an element of value is impacted (i.e., CQA fails to meet specification). Exploring risks requires that you understand what hazard(s) could occur that may affect a valuable element and what harm (s) could result. Risk is not determined by only looking at harms or hazards individually; it is the combination of both factors that allow us to determine risk. The measurement of the overall risk must also consider the mechanism in place to reduce the likelihood at which the hazard will result in harm. These mechanisms are the controls built into the process that lessen the likelihood of a failure or which detect a failure before they cause harm. Fig. 3.3 demonstrates the relationship between hazard, harm, and controls.

It is not uncommon for individuals to assess a risk-based purely on the potential impact of a failure without the consideration of the

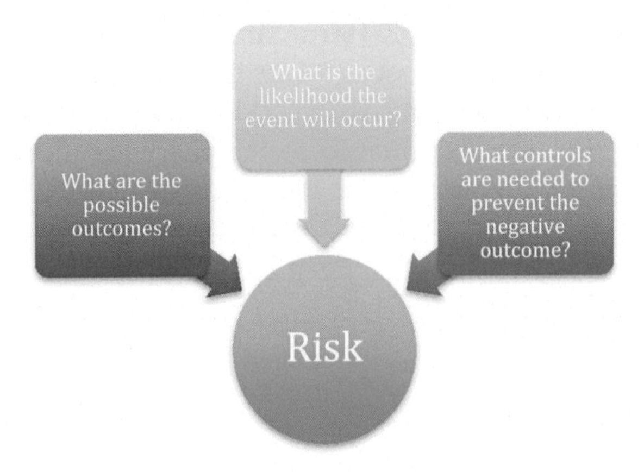

Figure 3.3 Risk: hazard, harm, and controls.

controls in place or the likelihood of an event occurring. In these cases, it is critical that the Facilitator breakdown the components to ensure that a multidimensional approach is taken and remains consistent through the assessment.

Risk Management: The Fine Print

One of the elements not present in the strict definitions of QRM and a commonly undervalued step of the life cycle is the initiation phase. This step in QRM process is instrumental in the developing a robust road map for executing a risk management activity. The initiation phase of risk management, while not revisited during the life cycle, is the foundation upon which our risk management activities stand and requires careful attention in order to ensure that the risk activity is well founded and mature.

QRM initiation can have various mechanisms, which are dependent upon the design and maturity of the QRM program. However, the following are critical elements that must be included and considered in the development of all risk activities:

- Risk Team Membership;
- Scope and Boundaries;
- Exploring the Process Map;
- The Risk Question and Tool Selection; and
- Scales for Assessing Risk

Risk Team Membership

Before beginning a risk assessment activity, the members of the risk team are carefully considered and must be a multidisciplinary group. A primary role within the team is the risk Facilitator. The risk Facilitator has an advanced knowledge risk management principles, risk management tools available for use, and an understanding of how to manage bias and heuristics within the risk team (Technical Report). Additional team membership includes persons with subject matter expertise in the process/product under review and a member of the quality organization. It is advised that the group of individuals participating in any given risk assessment session does not exceed six persons. If it is determined that more than six team members are required, the Facilitator will determine when it is necessary to have the entire team present and when it is optimal to include only a small portion of the group. The reason for keeping the risk team small is to ensure that each team member has the opportunity to give voice to the risk assessment and to develop a cohesive team dynamic throughout the process (Bligh et al., 2006).

Goal, Scope, and Boundaries

Before embarking on the risk assessment step of the life cycle, it is critical that the team is aligned with respect to goal, scope, and boundaries of the activity. The goal of the risk assessment is defined as an objective of the exercise and details the desired outcome of the risk assessment. For example, a risk assessment may have the goal of defining equipment maintenance or calibration schedules or could be identify risks related to a particular process step. Understanding what the intended actions required or what is expected to happen as a result of the risk assessment will provide a pathway for developing the scope and boundaries of the assessment. When a complex risk assessment activity is assigned, the notion of where to begin and end the assessment can become overwhelming. The best way to overcome this obstacle is by determining exactly what is in scope of the assessment and where within that scope the team is required stay (boundaries). When a complex risk assessment or a risk activity across multiple process steps is required, brainstorming with team during the initiation phase will inform the team of where their effort is required. In some cases, the risk assessment goal, scope, and boundaries are straightforward. For example, if the team is tasked with assessing the compliance risks related to the labeling of single product within at one manufacturing plant. However, all too often, risk activities seek to explore multiple dimensions of many process steps and require numerous considerations when determining the scope and boundaries. For example, a team may be tasked with understanding the determining detectability and control of bioburden, endotoxin, and viral contamination across the upstream and downstream manufacturing of multiple products or for performing an assessment of a manufacturing process from beginning to end (end-to-end assessment). In this case, the team needs to explore what kinds of failures are relevant to the risk assessment such as equipment, material, process flow, or environment and the strategy for performing the assessment such that all elements are evaluated with equal rigor. Given the scope of the assessment, a series of potential failures may be relevant or only a portion is considered within the assessment. A visual aid, such as a process map (described below), can provide the team with the structure needed to stay within the scope and boundaries of an assessment.

Exploring the Process Map

The risk team will be best prepared when a current process map is available during the initiation phase of the risk activity. Process maps assist in visualizing where the risk assessment will begin and end and clearly define the process steps within the scope of the risk assessment. By visualizing each step and understanding the relationships between those steps, the team can easily define the boundaries of their activity. In the event that the process under review does not have current or well-defined process map, the team should create one in order to facilitate the risk assessment activity to facilitate a well-organized effort. Additionally, process maps are critical in outlining the architecture of the documented assessment itself.

The Risk Question and Tool Selection

During the initiation phase, it is critical to define the intent of the risk assessment. This is achieved through constructing a Risk Question. A Risk Question is the question that the outcome of the risk assessment seeks to answer or the goal of the risk activity. The creation of the Risk Question for an assessment should take into account the scope and boundaries of the risk assessment and provide context on the goal of the assessment. Examples of risk questions are provided in Table 3.1.

The scope and boundaries of the risk assessment should be evident within the Risk Question and referenced by the team throughout the risk assessment to ensure that they are fulfilling the objective of the

Table 3.1 Example Risk Questions.

Topic	Potential Risk Question
Environmental monitoring	What are the microbial and particulate risks associated with the classified areas of the Raleigh Facility?
Process validation	What are the quality and compliance risks of performing two validation runs of Product CDMC?
Facility	What are the quality risks of building a new facility in Southern Texas?
Technical transfer	What the compliance risks of transferring the manufacture product A from Site 21 (Brussels, Belgium) to Site 67 (San Francisco, CA)?
In process testing	What are the microbial risks during downstream manufacturing of Product MEB?

activity. It is important to recognize that a Risk Question may be updated overtime such that it is an accurate reflection of the stated goal of the risk activity. Visualizing the "Line of Sight" (Baseman et al., 2016) will link the risk question with the ultimate goal of the assessment.

Once a Risk Question is composed and agreed to with the team, the risk management tool(s) to accomplish the goal of the assessment is selected. ICH Q9 describes the number of tools that are available to the industry and examples of when to use these tools. ICH Q9 states "the level of effort, formality and documentation of the QRM process should be commensurate with the level of risk" (ICH Q9, 2006). This principle allows us the opportunity to, based upon our Risk Question and objective, select the tool that is most appropriate to fulfill the needs of the assessment. Commonly used risk management tools are listed below in order of formality:

- *Failure Modes and Effects Analysis (FMEA)*: Provides an evaluation of potential failure modes for processes and their likely effect on outcomes and/or product performance. FMEA relies on product and process understanding. FMEA methodically breaks down the analysis of complex processes into manageable steps. It is a powerful tool for summarizing the important modes of failure, factors causing these failures and the likely effects of these failures (ICH Q9, 2006). FMEA can be used for end-to-end process assessment to understand the failure modes, failure causes, and effects of the overall parenteral manufacturing process. Investment in this method will allow for better process understanding that results in a more comprehensive evaluation and prioritization of the risks identified, controls required, and validation studies needed.

- *Process Validation use*: Process mapping, validation program master planning.
- *Hazard Analysis and Critical Control Points (HACCP)*: Systematic preventive approach to biological, chemical, and physical hazards in production processes that can cause the finished product to be unsafe, and designs measurements to reduce these risks to a safe level. HACCP identifies critical control points (CCPs) to inform a monitoring program.

HACCP can be used to determine in critical areas required for monitoring and control. Process validation studies should then be designed to address the effectiveness of the control of these critical areas.

- *Process Validation use*: Validation study scope determination.
- *Preliminary Hazard Analysis (PHA)*: A tool of analysis based on applying prior experience or knowledge of a hazard or failure to identify future hazards, hazardous situations, and events that might cause harm, as well as to estimate their probability of occurrence for a given activity, facility, product, or system. The tool consists of: (1) the identification of the possibilities that the risk event happens, (2) the qualitative evaluation of the extent of possible injury or damage to health that could result, (3) a relative ranking of the hazard using a combination of severity and likelihood of occurrence, and (4) the identification of possible remedial measures (ICH Q9, 2006). PHA has the most robustness when there is not a lot of historical data available to inform the risk assessment and may be used to determine the level of validation required.

- *Process Validation use*: New process study design.
- *Risk Ranking and Filtering*: A tool for comparing and ranking risks. Risk ranking of complex systems typically involves evaluation of multiple diverse quantitative and qualitative factors for each risk. The tool involves breaking down a basic risk question into as many components as needed to capture factors involved in the risk. These factors are combined into a single relative risk score that can then be used for ranking risks. "Filters," in the form of weighting factors or cut-offs for risk scores, can be used to scale or fit the risk ranking to management or policy objectives (ICH Q9, 2006). Risk ranking can be particularly useful in prioritizing validation study efforts, determining the effect of process changes, the need and extent for additional validation studies, and effective validation acceptance criteria.

- *Process Validation use*: Acceptance criteria determination.
- *Fault Tree Analysis (FTA)*: An approach that assumes failure of the functionality of a product or process. This tool evaluates system (or subsystem) failures one at a time but can combine multiple causes of failure by identifying causal chains. The results are represented

pictorially in the form of a tree of fault modes. At each level in the tree, combinations of fault modes are described with logical operators (AND, OR, etc.). FTA relies on the experts' process understanding to identify causal factors (ICH Q9, 2006). FTA can be useful in determining the effect of parenteral manufacturing process steps, conditions, or changes, the critical nature of such steps, conditions, or changes, and their potential effect on process performance and product quality. These aspects can then be used to design validation studies.

- *Process Validation use*: CAPA and change control evaluation.
- *Checklist*: A way to systematically analyze groups of data against a predetermined list of criteria or to determine impact of hazards to things of value. Checklists can be used to help organize, match and evaluate process and validation data to determine the effectiveness and outcome of the validation study.

- *Process Validation use*: Evaluation of study results.

Each risk tool has its own strengths and weaknesses, and carefully considered when deciding the tool to use. The structure, formality, and intent of the tools vary and choosing an inappropriate tool may not yield the desired outcome of the assessment. In order to ensure that you have chosen the appropriate tool for your risk assessment, consider the goal or what you hope to learn from the activity of your risk assessment and determine if the elements explored through a recognized tool are appropriate to meet those needs. To determine which tool is appropriate, it may be helpful to test a few different strategies against multiple tools to determine if the outputs are reflective of the desired outcome. When the appropriate risk tool is not readily apparent, there are considerations that the team can explore in order to determine which tool is best fit for the described task. Evaluating the questions listed below will assist narrowing down the best options.

Questions to consider when selecting a risk tool:

- Is the assessment of a new system, process or product?
- Is the system, product, or process a high-risk or high-impact item?
- Is the system process or product poorly understood?
- Is there enough available information on design and processing parameters to warrant the use of an extensive risk tool?

It is important to note that, while FMEA has become the most common risk tool executed by our industry, it is by no means the appropriate tool to use in all circumstances. If after reviewing the relevance of the recognized risk tools the team determines that needs of the risk activity will not be met with one of the available tools, it is not uncommon to use the principles of risk management to create a custom approach. This can be particularly instrumental when the parameters to be explored through an assessment are elements other than likelihood and severity.

Scales for Assessing Risk

Before beginning the risk assessment, the risk team will define the criteria used to evaluate the risks. Depending upon the risk tool deemed appropriate to answer the risk question, these criteria may be either qualitative or qualitative. FMEA is one tool where quantitative elements of the components of risk are scored whereas the HACCP utilities a qualitative scoring mechanism to determine overall risk. In many circumstances, defining qualitative and quantitative risk scores can be riddled with misinterpretation, debate among team members, and inconsistent application of risk levels. To overcome this, a Keyword approach to the creation of risk ranking criteria is encouraged. The Keyword approach assigns specific words to risk ranking criteria that are meaningful to both the team members and the objective of the risk exercise (Baseman et al., 2016). Developing a set of Keywords when executing a risk assessment will add an additional layer of ownership amongst the team members in answering their risk question and reduces the likelihood of scoring debate and confusion at the time of scoring the risk elements.

For most risk assessment methods, risk-ranking criteria will be defined for severity and likelihood of occurrence. In the case of FMEA and/or custom tool applications, detectability will require risk-ranking criteria.

Table 3.2 provides a side-by-side comparison of risk ranking criteria for assigning severity to materials used in a filling operation where general terms are used and where a keyword approach is applied. This example shows a 5-point scale; however, 3- or 10-point scales may also be used (Table 3.3).

After the risk ranking criteria are defined, it is important that the team come to agreement on the actions that are required because of

Table 3.2 Example Severity Criteria Comparing Generic Criteria With a Keyword Approach.

Ranking	Example Generic Criteria	Example Keyword Criteria
Negligible	No impact to product quality	Only applicable to the filling process step when the vials are stoppered
Minor	Minor impact to product quality	Material Transfer—non product contact materials
Serious	Moderate impact to product quality	Material Transfer—product contact materials
Critical	Potentially significant impact to product quality	Filler set up—non product contact materials
Catastrophic	Significant impact to product quality. Represents a risk to patient safety	Filler set up—product contact materials

Table 3.3 Example Likelihood of Occurrence Criteria Comparing Generic Criteria With a Keyword Approach.

Ranking	Example Generic Criteria	Example Keyword Criteria
Frequent	Failure almost inevitable	More than one occurrence per day or
		Greater than 1 occurrence in 10 opportunities
Likely	Failure likely	One occurrence per week or
		Greater than 1 occurrences per 100 opportunities
Occasional	Failure moderately likely	One occurrence every 1–6 months or
		One occurrence in 10,000 events
Unlikely	Failure unlikely	One occurrence per year or
		One occurrence in 100,000 events
Remote	Remote likelihood of failure	One occurrence every 1–5 + years or
		One occurrence in 1,000,000 events

the risk ranking. Typically, three buckets result from a risk assessment: High, Medium, and Low Risk which may be result of a Risk Estimation Matrix (REM) (Fig. 3.4) or by defining numerical thresholds (i.e., risk priority numbers). When utilizing the REM, it is the intersection of each component of risk that determines the overall risk level for hazards identified.

Before starting the assessment actions required for each of the risk levels are defined in an action table. These thresholds are outlined to ensure a consistent approach to addressing all risks identified. Table 3.4 outlines an example of a risk action table.

		Severity				
		Negligible	Minor	Serious	Critical	Catastrophic
Occurrence	Frequency	Medium	High	High	High	High
	Likely	Low	Medium	High	High	High
	Occasional	Low	Low	Medium	High	High
	Unlikely	Low	Low	Low	Medium	High
	Remote	Low	Low	Low	Low	Medium

Figure 3.4 Example risk estimation matrix for determining overall risk.

Table 3.4 Criteria for Risk Control and Acceptance.

Overall Risk Level	Required Action/Acceptability of Residual Risk
High	Mitigation required
	Residual risk is unacceptable; further mitigation and approval by the site decision makers is necessary to accept residual risk
Medium	Mitigation required, unless appropriate justification is provided
	Residual risk may be acceptable, provided appropriate justification is documented and approved by the site decision makers
Low	No further action required
	Residual risk is acceptable

High Risk elements are generally considered unacceptable and require mitigation, Medium Risks are considered acceptable only with a scientific rationale present to justify the risk, and finally Low Risk items are broadly acceptable.

Quality Risk Management Life Cycle

Risk Assessment

Once the initiation phase of the QRM process is complete, the risk assessment portion of the life cycle begins. There are three phases of the risk assessment step: Risk Identification, Risk Analysis, and Risk Evaluation. The first phase, risk identification, is the step in which hazards are identified. This is the portion of the risk assessment when all hazards within the scope of the risk assessment are documented by answering the question "What can go wrong?" If the team struggles in determining hazards or the hazards are not readily apparent, it can be helpful to perform a fishbone diagram (Ishikawa) to identify hazards. Additionally, mind maps and affinity diagrams are also helpful in hazard identification. During the risk identification phase, causes attributed to each hazard are documented and the resultant harms of the hazards. Once each of the elements risk is identified, risk analysis is performed. Risk analysis is the portion of the assessment where the risks are ranked and the criteria developed during the initiation phase (i.e., keywords for likelihood and severity) are applied to determine the overall level of risk. The final portion of the risk assessment is risk evaluation. In this phase, the team compares the overall risk level (as determined in risk analysis) to the action table designed in the initiation stage of the risk management process to determine if the overall risk level is acceptable or if controls are needed to reduce the risk to an acceptable level.

Risk Control

The risk control phase of the QRM process is composed of two elements: risk reduction and risk acceptance. Risk reduction focuses on reducing the identified risks to an acceptable level. For example, if it were determined through historical that a CQA is impacted by a repeated equipment failure, the risk reduction strategy would be to minimize or eliminate the likelihood that the offending equipment will fail. When considering a strategy for reducing the overall risk level, a commercial manufacturing site has a better chance of reducing the likelihood of the failure and not the severity of that failure. This is because, in most circumstances, reducing the severity of a failure is not possible. For example, consider microbial contamination, in most circumstances, microbial contamination is a showstopper for a manufacturing facility and renders a product adulterated and considered a critical severity. Since it is not possible to make microbial contamination less critical, we focus on reducing the likelihood that microbial contamination will occur. In many cases, there is typically not one control that is able to prevent a catastrophic event but the combination of controls, which provides us with confidence that a state of control is maintained.

When selecting the risk mitigation or risk control strategy, the following options (listed in order of effectiveness) can be taken to reduce the risk level (ISO/IEC Guide 51, 1999; Vesper, 2006).

- Eliminate the hazard;
- Uncouple or loosely couple the process;
- Decrease the frequency of an event happening;
- Duplicate the assets;
- Diversify;
- Change the source of failure;
- Proceduralize; and
- Training

Once a risk control strategy is defined, the actions resulting from the control strategy must be taken before the risk can be rescored. This is an importance step in the QRM process and will be discussed in more detail in the risk review and monitoring steps.

Risk acceptance is defined as the decision to accept risk (ICH Q9, 2006). Decisions regarding whether to accept a risk or to mitigate risks are delegated to the organization's leadership team. Their responsibly for performing this function are clearly outlined in ICH Q9, Q10, and ISO 31000. Risk-based decision making plays a role in not only determining the level of action to take but also in establishing the organization's risk tolerance (i.e., the amount of risk the organization is willing to live with). In rare cases, it may not possible to reduce a high risk an acceptable level and when these circumstances are presented, they must be carefully considered and documented with a risk–benefit analysis. Acceptance of high risks is a decision to be made by your organization's leadership and is only acceptable when there is documented evidence that the benefits of the product outweigh the risk identified. In the case of medium risks, if the severity ranking is the risk element which is driving the overall risk score while the likelihood of occurrence has been reduced as low as possible, this risk may be accepted and justified.

Risk Communication

Risk communication is the sharing of risk information generated throughout the risk management process and can take various forms depending upon your organization's structure. One of the primary ways in which risk information is shared across an organization is through the use of a Risk Register. Risk Registers outline the most critical risks by site, process or product and allow for Top Management to identify the areas of the business that are most vulnerable. Risk reports generated as a result of risk exercises are also key communication elements. Risk reports will describe the risk management tool used to answer the risk question and provide a detailed description of each hazard assessed with justification of the risk ranking criteria selected. Regardless of the mechanism selected to best distribute risk information within your organization, it is important to ensure that all impacted parties are well informed of product/process/system risks and the strategies in place to prevent those risks from being realized. These parties include but are not limited to regulatory bodies, contract manufacturing organizations, other sites within your organization, and the patient. The frequency of communication and the factors that trigger risk communication should be defined within your organization.

Risk Review

The final step in the risk management process is review and monitoring. The intent of this stage is to update existing risk assessments to account for process changes and to incorporate knowledge gained since the assessment was initially performed. Risk review is critical for determining if risk reduction measures taken during the risk control phase were effective and for ensuring that the process maintains a well-controlled state. As described in ICH Q9, a risk review can either be planned (i.e., scheduled to occur periodically) or unplanned (i.e., trigger by an event such as a recall). In either case, elements to consider when performing risk review include: deviations or change controls initiated since the risk assessment was performed, changing regulatory requirements and results of the annual product review.

Stage 3 of Process Validation has the review portion of the life cycle built in. During this stage, data are reviewed periodically, and changes are made to account for any variable observed or trends identified overtime (Guidance for Industry, 2011). While the risk review portion of the life cycle can appear to be more of a documentation exercise, it is important to recognize that this is the step in the life cycle where it is possible to reflect and learn more about a new process or improve an existing process. When a risk assessment is initially performed, it may be difficult to estimate the frequency of failure until after the process has been run several times. Risk review provides the opportunity to input that data and take credit for what is performing as expected and to identify where refinement of the process is needed. The frequency of risk review is dependent upon both the criticality of the process examined during the risk assessment and also the maturity of the process examined. For example, if a new process is implemented, it may be appropriate to perform risk review annually in order to gather and refine the process. Once this process has reached a steady and predictable state (i.e., fewer deviations and changes to the process overtime), the frequency of review can be scaled back.

Conclusion

One of the most valuable benefits of implementing risk management in your organization is the ability reduce uncertainty in the manufacturing process. Risk management tools provide data driven mechanism to evaluate where knowledge is needed and also provide a consistent structure allowing for knowledge management. The key to successful a risk management program lies not only in the application of principles and adhering to the QRM life cycle but also in the culture of the organization. Implementing a procedure or policy on how to perform risk management is not enough to reap the benefits that QRM has to offer. The success of a program its benefits (i.e., increased productivity, reduction of failures and anticipation of failures before they happen) can only be achieved when the program is driven from the top down. This is to say that our organization's leadership must not only support the QRM programs with qualified resources but also begin using the outcomes of risk assessments for risk-based decision making and strategic planning. A large part of moving an organization from simply documenting risk assessment to meet regulatory needs to utilizing them to increase knowledge and reduce uncertainty is through recognizing that not all risk assessments will yield low risk outcomes and embracing the high risks as opportunities for improvement.

References

AS/NZS ISO, 2004. 31000: 2009 Risk Management—Principles and Guidelines, third ed.

Baseman, H., Hardiman, M., Henkels, W., Long, M., 2016. A Line of Sight Approach for Assessing Aseptic Processing Risk: Part I, ValSource, May 31, PDA Letter.

Bligh, M.C., Pearce, C.L., Kohles, J.C., 2006. The importance of self- and shared leadership in team based knowledge work A meso-level model of leadership dynamics. J. Manager. Psychol. 21 (4), 296–318.

Guidance for Industry, 2011. Process Validation: General Principles and Practices. U.S. Department of Health and Human Services Food and Drug Administration Center for Drug Evaluation and Research (CDER) Center for Biologics Evaluation and Research (CBER) Center forVeterinary Medicine (CVM): January.

ICH Q10, 2000. Q10 Pharmaceutical Quality System; International Conference on Harmonization. <www.ich.org>.

ICH Q8, 2009. Pharmaceutical Development Q8 (R2); International Conference on Harmonization. <www.ich.org>.

ICH Q9, 2006. *Quality Risk Management* ICH Quality Guideline Q9: Quality Risk Management; International Conference on Harmonization. <www.ich.org>.

ISO 14971, 2009. Medical Devices—Application of Risk Management to Medical Devices, 5 January 2013, *ICH Q8*, PHARMACEUTICAL DEVELOPMENT Q8(R2); International Conference on Harmonization. <www.ich.org>.

ISO/IEC Guide 51,1999. - Safety Aspects - Guideline for their inclusion in standards.

Technical Report No. 54 Implementation of Quality Risk Management for Pharmaceutical and Biotechnology Manufacturing Operations.

Vesper, J.L., 2006. Risk Assessment and Risk Management in the Pharmaceutical Industry: Clear and Simple Hardcover.

CHAPTER 4

Equipment Cleaning Process

Igor Gorsky
Senior Consultant, ConcordiaValsource LLC., Downingtown, PA, United States

The primary aim of this chapter is to outline the significance of an *Equipment Cleaning* as a *Process* (Gorsky, 2014a,b; Gorsky et al., 2016). A typical continuum of an *Equipment Cleaning Process* for Parenteral Products consists of:

- Development (if it exists) of Cleaning Procedures by Research and Development groups;
- Validation of Cleaning Procedures by Technical or Quality groups; and
- Application of Cleaning Processes by Operations/Manufacturing organizations.

Although most of the Cleaning Process and Validation written procedures typically state that changes are covered by a change management, these procedures are rarely changed. A brief description above constitutes usual continuum for hundreds of parenteral drugs' pharmaceutical companies. However, changes are seldom pursued as most of the companies perceive them as infringement upon "validated process." Therefore often these processes are "set in stone" and those that implementing them mainly disregard the rules of a process shift (Best and Neuhauser, 2006). Most of the time despite extensive global regulations and voluminous technical literature library, the subject, a failure of a Cleaning Process is not a "human error" but a failure to recognize Cleaning as a Process and treat it as such. Instead of pursuit of process understanding, as described in this book for a manufacturing processes, these failures cause organizations to perceive Cleaning Process as a "necessary evil," which they need to tolerate mostly due to regulations. Many would even try to avoid it altogether and argue that it is not a value-added activity. However it is obvious that this kind "philosophy" is wrong, and Cleaning must be recognized as an

Principles of Parenteral Solution Validation. DOI: https://doi.org/10.1016/B978-0-12-809412-9.00003-4

important and necessary part of parenteral manufacturing, especially taking into consideration a route of administration of products. If one does not develop an appropriate method for cleaning process residuals, does not appropriately validate this method, and does not assure that it is still viable throughout the life cycle of the product, one is adding another unknown variable into an actual manufacturing process, and we all know that "uncontrolled variation is the enemy of quality."[1]

Therefore this chapter should help practitioners in development, utilization, and maintenance of Cleaning Validation programs so that they can reduce process variability. To achieve this goal, we will touch upon an important aspect of Cleaning Validation that gained traction in the last several years due to implementation of Risk-based Life Cycle approach to Cleaning Validation (ASTM E3601-18e) built on principals of ICH Q8, 9, 10 Guidance, FDA Guidance for Industry: Process Validation, as well as to as well as EU Annex 15: Qualification and Validation. This subject is development of limits for the process residue. Knowledge of the limits setting strategies should help in gauging one's Cleaning Validation program.

Establishing Limits

The subject of soil residue limits setting is of an utmost importance for the obvious reason—it is a measure of the cleaning process effectiveness and consistency. We measure success of a Cleaning Validation study by assuring that we meet predetermined criteria based on removal of the soil to a level below an established limit. Therefore establishing limits is one of the pivotal steps in Cleaning Validation continuum. Although there are many sources on this topic and we can specifically mention a few excellent references [Parenteral Drug Association (PDA), 2012; Walsh, 2011] for various methods of setting soil residual limits as well as history of the subject, our goal would be to inject employment of science and knowledge into this exercise. First we will briefly talk about Health-Based Exposure Limits, since this subject has been debated for a few years and most definitely benefit from some level of demystification. We will only summarize few important points to consider when evaluating such limits. There are several major terms being currently used globally for these limits.

[1] Attributed to Edward Deming (1980) in Kang and Kvam (2012).

They are listed in the Table 4.1 along with their respective sources and some notes that describe their intentions and use.

It is important to note that the first two terms—ADE and PDE are the most referenced by the regulators. However if the first term—an

Table 4.1 Health-Based Exposure Limits Terms

Term	Description	Source
Acceptable Daily Exposure	Acceptable Daily Exposure or ADE represents a dose that is unlikely to cause an adverse effect if an individual is exposed, by any route (e.g., intrathecal, inhaled), at or below this dose every day for a lifetime	ISPE Risk-MaPP (Risk-Based Manufacture of Pharmaceutical Products), 2010
	ADEs is typically based on NO[A]EL (No Observable Adverse Effect Level) usually expressed in mg/kg body weight/day	
Permitted Daily Exposure	Permitted Daily Exposure or PDE is defined in the present guideline as a pharmaceutically acceptable intake of residual solvents to avoid confusion of differing values for ADI's of the same substance	ICH Q3C R5, Impurities: Guideline for Residual Solvents
	The PDE represents a substance-specific dose that is unlikely to cause an adverse effect if an individual is exposed at or below this dose every day for a lifetime. It is based on NOEL (No Observable Effect Level), referenced in ICH Q3C Residual Solvents and in EudraLex Annex 15: Qualification and Validation Draft Guidance	
Acceptable Daily Intake	Acceptable Daily Intake or ADI is a measure of the amount of a specific substance (originally applied for a food additive, later also for a residue of a veterinary drug or pesticide) in food or drinking water that can be ingested (orally) on a daily basis over a lifetime without an appreciable health risk. ADIs are expressed usually in mg/kg body weight/day	WHO, 1987 "Principles for the safety assessment of food additives and contaminants in food"
Threshold for Toxicological Concern	A Threshold of Toxicological Concern or TTC has been developed to define a common exposure level for any unstudied chemical that will not pose a risk of significant carcinogenicity or other toxic effects. TTC value of 1.5 μg/day intake of a genotoxic impurity is considered to be associated with an acceptable risk (excess cancer risk of <1 in 100,000 over a lifetime) for most pharmaceuticals. From this threshold value, a permitted level in the active substance can be calculated based on the expected daily dose	EMA, Guideline on The Limits of Genotoxic Impurities, 2006

Acceptable Daily Exposure (ADE) has been routinely used in the United States, the second term Permitted Daily Exposure (PDE) is been consistently referenced by the European Guidance. These two terms although similar have some differences, at least in the eyes of some regulators. ADE is extrapolated from NO[A]EL while PDE is typically based on extrapolation from NOEL. The difference between NO[A]EL and NOEL is very important. The Table 4.2 compares description for these two levels.

Another reference to consider with regards to the difference between two levels is FDA's view which was cited with respect to comments raised by industry for the "Estimating the Maximum Safe Starting Dose in Initial Clinical Trials for Therapeutics in Adult Healthy Volunteers" Guidance for Industry. In their response, FDA said that "the NO[A]EL is not the same as the no observed effect level (NOEL), which refers to any effect, not just an adverse one, although in some cases the two might be identical" (FDA Guidance for Industry, 2006). Furthermore "the definition of the NO[A]EL, in contrast to that of the NOEL, reflects the view that some effects observed in the animal may be acceptable pharmacodynamics actions of the therapeutic and may not raise a safety concern."

Cleaning Validation practitioners should consider these differences and utilize a toxicologist or a person with adequate training in pharmacology and toxicology to develop and document Health-Based Exposure Limits Assessments for soil residual limit calculations.

Although undisputedly important, the Health-Based Exposure Limits should not be the only ones utilized for measuring of a Cleaning Process. In order to gain a thorough understanding of the

Table 4.2 NOEL versus NO[A]EL

NOEL	NO[A]EL
No Observed Effect Level for pharmaceuticals is currently based exclusively on an extrapolation of the results of animal toxicity studies	No Observed Adverse Effect Level denotes the level of exposure of an organism, found by experiment or observation, at which there is no biologically or statistically significant (e.g., alteration of morphology, functional capacity, growth, development or life span) increase in the frequency or severity of any adverse effects in the exposed population when compared to its appropriate control

Table 4.3 Risk Levels for Setting Cleaning Validation Limits

sould	Limit Term	Description
Level 1	Health-Based Exposure Limits	A risk-based approach that reviews and evaluates pharmacological and toxicological data of individual active substances, and thus enable determination of patient safe threshold levels as referred to in the GMP guideline which determines the level of containment and dedication of facility and equipment to be used during manufacturing and cleaning of the drug product and drug substance
Level 2	Visible Limits	This a quantitative visual detection limits. A visual detection limit under specified viewing conditions can be determined by spiking coupons of the equipment surface materials with solutions of the different levels of the soil residue. Then a panel of trained observers determines the lowest level at which residues are clearly visible across the analyzed surface. This quantitatively defined level is the limit at which visual inspection could be conducted. Typical values reported in the literature for a visual detection or a visible limit is below or equal to 4 $\mu g/cm^2$
Level 3	Capability Analyses Limits	Developed from typical Cleaning Validation Limits such as: • no greater than 1/1000th of the lowest clinical dose of the soil residual in the maximum daily dosage of the next product to be manufactured, or • a maximum of 1–10 ppm of the previous active substance in the next product manufactured Whichever of these criteria results in the lowest carryover is a typical limit applied for cleaning validationAlthough, these "limits do not take account the available pharmacological and toxicological data and may be too restrictive or not restrictive enough" with a knowledge of Health-Based Exposure Limits and Visible Limits, they could be practically utilized as an Alert and/or and Action limits for Cleaning Validation studies especially in determining through a statistical process capability analysis an actual level of the cleaning process

level of cleanliness, it is recommended to utilize the limits setting practice using multiple levels. These levels are summarized in Table 4.3.

Usually, for pharmaceutical drug products Health-Based Exposure Limits would be the largest unless there are toxicological hazards associated with an Active Pharmaceutical Ingredient (API). The Visible Limits are typically very similar to historically used limits. An illustration of this relationship is shown in the Fig. 4.1 below:

It is important to note that MACO 1 ppm limits are readily achievable by most validated cleaning processes. In addition, these cleaning processes are typically found to be statistically capable. Therefore it is highly helpful to use statistical process capability analyses for evaluation of these results[2].

[2] To use statistical analysis, utilize same precautions as referenced in FDA Guidance for Industry: Process Validation: General Principles and Practices, January 2011, p. 14.

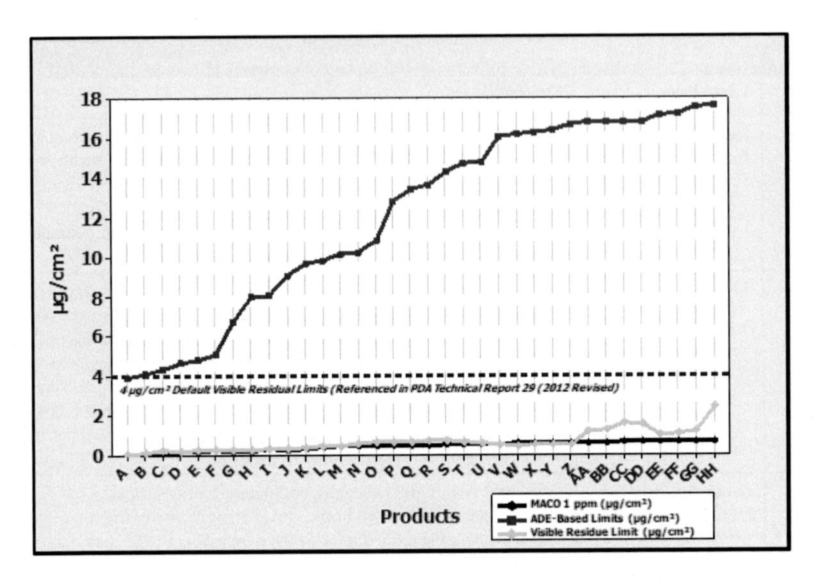

Figure 4.1 Typical visible residual limits, MACO limits (Cleaning Validation Achievable) and health-based exposure limits.

We will discuss several tools that may be utilized for the measurement of equipment cleaning process (ECP) during all three stages of its continuum:

- Development of Cleaning Procedures by Research and Development groups (if it exists);
- Validation of Cleaning Procedures by Technical or Quality groups; and
- Application of Cleaning Processes by Operations/Manufacturing organizations.

The key to control for any process resides in the understanding of its variability. How much do we know about the variation of our processes or systems and are we in the state of control? These questions require answers to achieve success in the execution of ECP. Understanding, detection, response, and control, from input through output of variation, consumed a focus of the revised FDA Process Validation Guidance (FDA Guidance for Industry, 2011) to Industry (Figure 1 in Chapter 1: Process Validation: Design and Planning[3]). So how does one go about control of variation? To do that one has to

[3] Figure adopted from Baseman (2013).

enable a Risk Management System[4]. To be highly effective, a Risk Management System needs to be initiated early on during the design and development of ECP. A rigorous application of Risk Management tools during Stage 1 (Process Design Stage) will help the practitioner to assess, understand, and ultimately, control the level of variation in systems and processes. Therefore, Critical Quality Attributes and Critical Process Parameters should be established during risk assessment exercises. In addition to Risk Assessment exercises, experiments should be conducted to attain data about cleaning process being developed and understood. These Stage 1—Process Design studies should preferably utilize a statistical Design of Experiments (DoE) where appropriate. DoE is "a structured, organized method for determining the relationship between factors affecting a process and the output of that process" (ICH Quality, 2009).

As stated in the FDA Process Validation Guidance, "risk analysis tools can be used to screen potential variables for DoE studies to minimize the total number of experiments conducted while maximizing knowledge gained." Further this guidance says that "the results of DoE studies can provide justification for establishing ranges of incoming component quality, equipment parameters, and in-process material quality attributes" (ICH Quality, 2009). At the time of the Process Design, it is generally recognized that not all possible sources of variability will be known; however, if risk management is exercised to develop insightful DoE studies, they should help developing low variability process.

Additionally, comparative cleanability studies could be performed to compare products. For instance, Fig. 4.3 shows Total Organic Carbon (TOC) results from such a study for four products. The study, the results of which are illustrated in Fig. 4.3, was performed side-by-side by spiking coupons with products' residuals and cleaning them using a worst-case cleaning method used for all four products. The time taken to remove residuals was measured and compared. Prior to the experiment, all four products were evaluated by applying the firm's theoretical "Worst Case" product matrix. Companies often use these matrices to determine the worst-case product which ultimately will be

[4] Technical Report No. 60: Process Validation: A Lifecycle Approach; Parenteral Drug Association: 2013, www.pda.org/bookstore (accessed 14.03.13).

used for a cleaning validation study. Typical evaluation factors include (but are not limited to):

- Solubility of active;
- cleanability of the active concentration;
- toxicity; and
- complexity of the cleaning procedure.

The first two products (Product A and Product B) during this theoretical evaluation showed very similar scores. It should be noted here that API used in these products were very similar compounds, had similar solubility, and therefore were expected to show similar results for their cleanability. However, as seen in Fig. 4.2, the study results were strikingly different. In order to conduct a side-by-side evaluation, the data for all products were normalized. This example shows how important it is to evaluate and compare all of the parameters by normalizing the data and plotting them on the same plane. What Fig. 4.3 shows is that although Product B contains almost twice the concentration of an API, which, as we said earlier was theoretically very similar in solubility to Product A, it took approximately half the time to clean, as was measured by visual observation. In this case, although initial evaluation of the products information did not reveal any significant differences, the results of the study and subsequent investigation

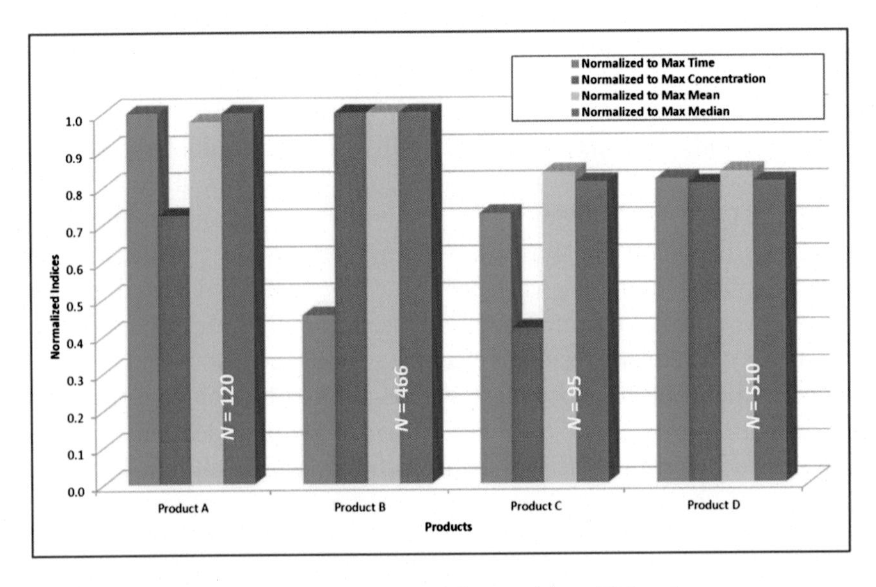

Figure 4.2 Cleanability study comparison and commercial cleaning validation TOC results.

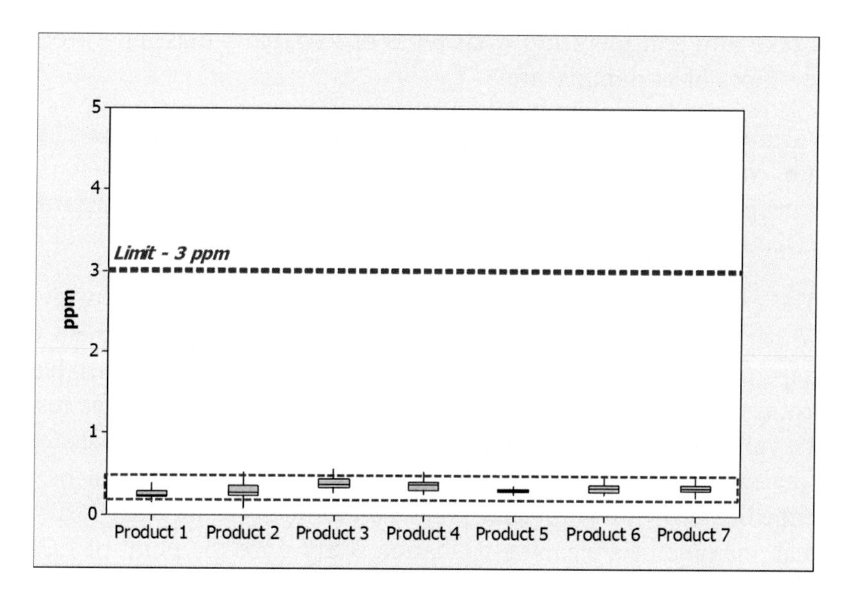

Figure 4.3 Comparison of TOC swabs for 7 products.

showed that chemical interactions of ingredients in the formulation of Product A significantly contributed to a longer time to remove residuals. This was important to learn because just a theoretical evaluation would not have revealed this complexity. Therefore it is recommended to performed studies and plot data normalizing them when appropriate so that they could be presented on the same plane to identify previously unknown sources of variability.

The second stage of validation is called Process Performance Qualification. This stage is customarily referred to as "Cleaning Validation". Usually, three consecutive successful runs are performed to qualify the process using a well-characterized, well-documented, and consistent cleaning procedure. During these studies, one cleans the equipment, collects appropriate samples, and evaluates the data using predefined statistical tools. We should mention here that for years, it was not habitual to use statistics to evaluate ECP, and it might be a new concept for many readers. The author of this chapter strongly encourages the usage of such methods as they will provide meaningful insights into sources of ECP variability. Remember that a result that you did not expect or is hard to explain still tells you a story. And we believe that each process tells a "story." So, what statistical methods should we employ during Stage 2—Process Qualification? We should first take into consideration what types of variability exist in a process. These types of variability are:

- Variability within each individual cleaning run (also called "inter-run" variability); and
- Variability between the cleaning runs (also known as "intraruns" variability).

The examination of variability within the cleaning run often reveals those parts of the equipment train or individual parts of equipment that perhaps are harder to clean or sample, thus possibly causing variable or aberrant results. However if the cleaning process is consistent, the results of the validation studies should illustrate this consistency. The first step in the review of the data would be checking its normality. Do not be discouraged, though, if your data set is nonnormal. Nonnormal is a very typical outcome of a cleaning validation study since the point of ECP is to completely remove manufacturing and cleaning process residuals. Therefore the results of many sampling and tests yield either "0" or close to "0." Upon finding out how normal your data are, one should calculate confidence intervals (CIs) around sample results population data set.

A CI gives an estimated range of values which is likely to include an unknown population parameter, the estimated range being calculated from a given set of sample data. If independent samples are taken repeatedly from the same population (consistent cleaning process should produce same results), and a CI calculated for each sample, then a certain percentage (confidence level) of the intervals will include the unknown population parameter. CIs are typically calculated so that this percentage is 95%, but we can produce 90%, 99%, 99.9%, etc., CIs for the unknown parameter. Then we can examine tolerance intervals[5] and perform early process capability analyses using normal and nonnormal capability analyses (depending on normality of the data sets). When data are found to be nonnormal, find the best fitting model.

The following intercleaning run variability analyses can be utilized for cleaning process qualification study examination:

- Individual value plot;
- box-plot;

[5] ISO16269-6 Statistical interpretation of data – Part 6: Determination of statistical tolerance Intervals.

```
Two-sample T for 0.5mg Run 1 vs 0.25mg Run 1

                N     Mean    StDev   SE Mean
0.5mg Run 1    12   0.0140   0.0288    0.0083
0.25mg Run 1   12    0.046    0.139     0.040

Difference = mu (0.5mg Run 1) - mu (0.25mg Run 1)
Estimate for difference:  -0.0318
95% CI for difference:  (-0.1222, 0.0585)
T-Test of difference = 0 (vs not =): T-Value = -0.78  P-Value = 0.454  DF = 11
```

Figure 4.4 Two-sample T-test and CI: Product A 0.5mg Run 1, Product A 0.25mg Run1.

- ANOVA;
- two-sample *t*-test;
- two one-sided *t*-test;
- nonparametric tests (e.g., Mann–Whitney test).

We will illustrate some of these tools in the following few examples. Fig. 4.3 shows a Box-plot of TOC results of the cleaning studies for seven products that utilize the same clean-in-place cycle.

As we can see in Fig. 4.3, all the cleaning studies are quite similar and are well within set limit of 3 ppm.[6]

Another test to evaluate the difference between cleaning validation runs is to use two sample *t*-tests to compare data populations for two runs. Fig. 4.5 shows the result of such tests as produced by statistical package Minitab.

We may inspect several parameters presented in Fig. 4.4 summary. However the most important one we are looking for is a *p*-value. The *p*-value or calculated probability is the estimated probability of rejecting the null hypothesis of a study question when that hypothesis is true. When *p*-value is above .05 when using it to test null hypothesis that both cleaning runs represent same sample populations (because ECP should be consistent), then we can say that two population means are the same with 95% confidence. In this case, the *p*-value is .454 which is well above a .05 *p*-value.

Additionally, two one-sided *t*-test (TOST) can be used to determine the equivalence of two data sets. When a regular *t*-test is used to conclude there *is* a substantial difference, we must observe a difference

[6] Default is a limit presented for illustration only. Limits for each cleaning process should be determined and evaluated separately.

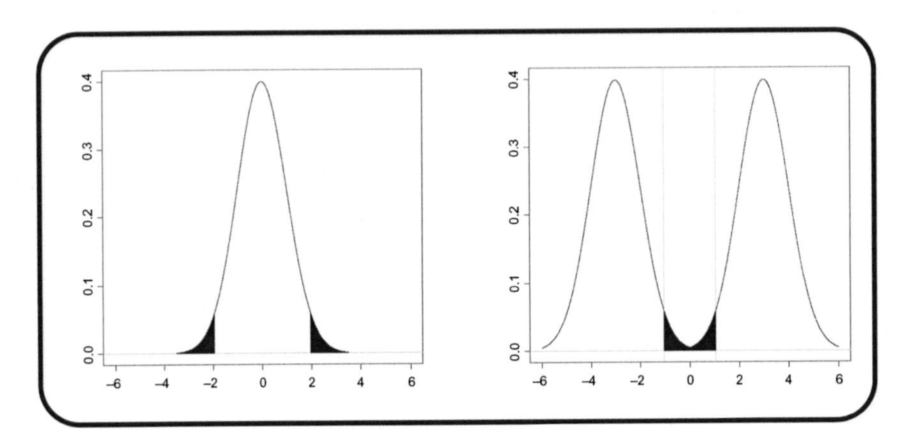

Figure 4.5 Regular t-test tails examination (left); *Two one sided t-test tails examination* (right).

large enough to conclude it is not due to sampling error, *p*-value above preset α (ex, .05). TOST applies with equivalence testing to conclude there *is not* a substantial difference, we must observe a difference small enough to reject that closeness is not due to sampling error from distributions centered on large effects (e.g., within 95%, 90%). Fig. 4.5 illustrates the difference in the population tail examination between a regular *t*-test and TOST.

TOSTs may be used to evaluate manufacturing process residual removal time for two Cleaning Validation runs or cleaning runs for two products using the same ECP. The swab tests' TOC results for two data sets may be analyzed using Minitab software. The prerequisite for this kind of test is an equivalent sample size for both populations and normality of both populations. Provided these prerequisites are met, Fig. 4.6 shows the outcome of this analysis for two products.

In the case of a cleaning evaluation (as seen in Fig. 4.7), this statistical approach has been applied to determine the relative cleanability of two products.

The TOST compares two group means and their CIs by comparing them to a predefined equivalence limit (e.g., 90%). The predefined equivalence limit is established by evaluating the variability involved with such evaluations. It should be noted that TOST is a statistical method accepted for evaluation of the equivalency between two groups of data such as bioequivalence (±20%) (FDA Guidance for industry, 2001) and Cleaning Validation (Chambers et al., 2005).

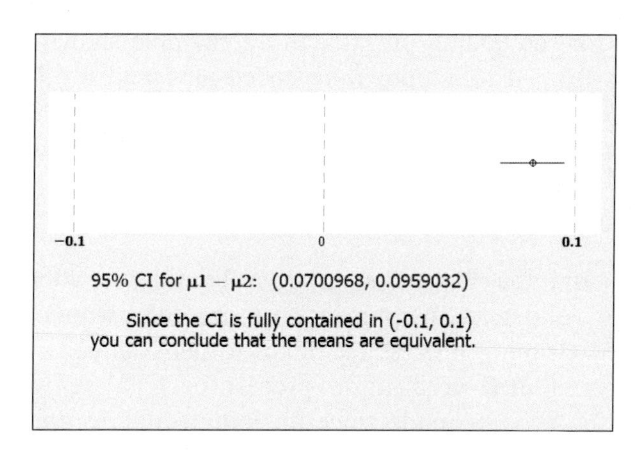

−0.1
0
0.1
95% CI for μ1 − μ2: (0.0700968, 0.0959032)
Since the CI is fully contained in (-0.1, 0.1)
you can conclude that the means are equivalent.

Figure 4.6 Equivalence test for Product A and Product B (90% confidence) (424 results).

```
                         N  Median
Product A_CODED       3065  1.0000
Product B_CODED        305  1.0000


Point estimate for ETA1-ETA2 is -0.0000
95.0 Percent CI for ETA1-ETA2 is (0.0000,-
0.0000)
W = 5582441.0
Test of ETA1 = ETA2 vs ETA1 not = ETA2 is
significant at 0.9103
```

Figure 4.7 Mann-Whitney test and CI: Product A_CODED, Product B_CODED.

In addition, if it is possible to pool samples from all runs, individual run charts can be plotted and overall ECP capability examined. One more example of intrarun variability is nonparametric analysis. Since population of TOC or other ECP results could be nonnormally distributed, a nonparametric test for medians could be used to compare cleaning validation swab test results. The Mann–Whitney test is such a nonparametric test of the null hypothesis that is used to show that two populations are the same against an alternative hypothesis, especially when a population tends to have larger values than the other.

This test has a greater efficiency than the *t*-test on nonnormal distributions, a mixture of normal distributions, and it is nearly as efficient as the *t*-test on normal distributions. Therefore it could be used to compare any distribution. Minitab software was used to perform the test. To perform this test certain procedures should be followed, particularly for the coding of data. In an example shown in Fig. 4.7, TOC results from 0 to 1 ppm were coded as 1 and results 1.1– 100 ppm were coded as 2. The Minitab summary shown in Fig. 4.7 illustrates that two populations are similar with 0.9103 (91.03%) significance, thus proving their equivalence from 0 to 1 ppm, which was appropriate for examined ECP.

The last and the longest stage of validation is Stage 3, which encompasses validation maintenance or in other words verification that qualified cleaning process continues to perform consistently. It is also called continued verification process or CPV. Obviously, any changes to the process made since an initial qualification should be evaluated to assess individual impact of each change as well as a cumulative impact of all changes. However it is also important to evaluate not only changes but also to establish meaningful verification program which would collect, evaluate, and trend data from each cleaning as well as conduct periodic cleaning verifications to assure that data attained during an original qualification study are representative of the cleaning process performed.

Typically, one should observe a process capability improve as manufacturing operation groups should become more proficient in performing qualified ECP. Fig. 4.8 illustrates 3 years' worth of continued verification studies.

As shown in Fig. 4.8, process capability indices of ECP improved from Ppk = 1.21 in Year 1 to Ppk = 2.17 in Year 2, and finally to Ppk = 2.43 in Year 3. One should notice that nonnormal process capability analyses calculations were used in these examples since the data sets were found to be nonnormal.

It also should be noted that the goal of CPV should not only be confirmation of qualified cleaning methods but more importantly optimizations of ECP through learning and knowledge management. Because the cleaning of equipment is a critical official process, it should be treated as such, and therefore be validated in the same way as a manufacturing process would be.

In conclusion, it is evident that a Process Validation life cycle approach with its use of statistics is applicable to ECP validation. The risk and knowledge management methodology help us understand our ESP process, the products of the processes (e.g., a clean surface), our

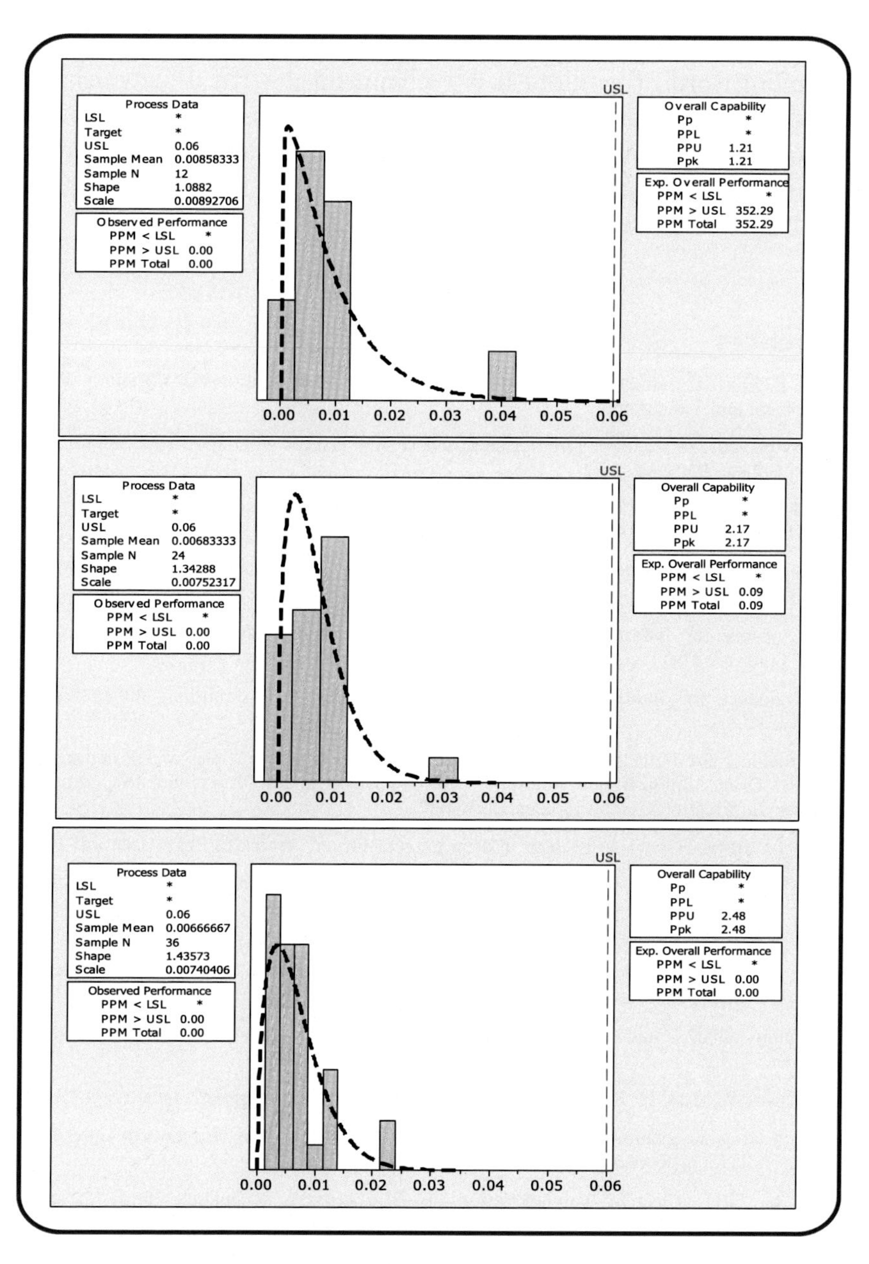

Figure 4.8 Process capability of Product A (0.5mg) CPV Year (1–3). Calculation based on Weibull distribution model.

variables, and ultimately gives us confidence during commercial production of parenteral pharmaceutical products. Finally, utilization of statistics and statistical process controls are essential to effective and efficient ECP post-performance qualification monitoring programs. One more word of caution: it is recommended "that a statistician or person with adequate training in statistical process control techniques develop the data collection plan and statistical methods and procedures used in measuring and evaluating process stability and process capability" (FDA Guidance for Industry, 2011).

References

ASTM E3601-18e Standard Guide for Science-Based and Risk-Based Cleaning Process Development and Validation.

Baseman H., 2013. Parenteral Drug Association (PDA) Process Validation Process Validation Training Course, Bethesda, MD.

Best, M., Neuhauser, D., 2006. Walter A Shewhart, 1924, and the Hawthorne factory. Qual. Safety Health Care 15, 142–143. Available from: https://doi.org/10.1136/qshc.2006.018093.

Chambers, D., et al., 2005. Analytical method equivalency: an acceptable analytical practice. Pharm Technol. 9, 64–80.

FDA Guidance for Industry, 2006. Estimating the Maximum Safe Starting Dose in Initial Clinical Trials for Therapeutics in Adult Healthy Volunteers.

FDA Guidance for Industry, 2001. Statistical Approaches to Establishing Bioequivalence. Rockville, MD.

FDA Guidance for Industry, 2011. Process Validation: General Principles and Practices, U.S. Food and Drug Administration: January www.fda.gov/downloads/Drugs/GuidanceCompliance-RegulatoryInformation/Guidances/ucm070336.pdf.

Gorsky, I., 2014a. How clean is clean in drug manufacturing? American Pharmaceutical Forum July/August.

Gorsky, I., 2014b. How clean is clean in drug manufacturing, part 2? American Pharmaceutical Forum November/December.

Gorsky, I., Hunff, C., Long, Dr. M., Hartman, J., 2016. How clean is cleaning validation level of maturity. American Pharmaceutical Forum January/February.

ICH Quality, 2009. Guideline Q8(R2) Pharmaceutical Development, Guidance for Industry, November.

Kang, Chang W., Paul H. Kvam, 2012. Basic Statistical Tools for Improving Quality. p. 19.

Parenteral Drug Association (PDA), 2012. Technical Report (TR) No. 29: Points to Consider for Cleaning Validation (Revised), PDA Publishing, Bethesda, December.

Walsh, A., 2011. Cleaning validation for the 21st century: acceptance limits for active Pharmaceutical ingredients (APIs): part I. Pharm. Eng. July/August.

Further Reading

Parenteral Drug Association, Technical Report 60 Process Validation: Lifecycle Approach.

CHAPTER 5

Quality Risk Management of Parenteral Process Validation, Part 2: A Risk-Based Quality Management System

Lori Richter
Senior Consultant, Valsource, Inc., Downingtown, PA, United States

Overview

Validation is an important element in an overall Quality Management System (QMS). It helps to demonstrate the reliability and performance of processes and support systems. Validation is linked and supported by all other QMS elements, including change control, audit, corrective actions, and deviation management. These QMS elements provide information as to the status and relative importance of given process steps and system aspects, essential to the proper design and evaluation of validation studies. Decisions should be made on a risk basis, ensuring that validation studies are designed to address those process steps and system aspects having the greatest impact of process performance. Likewise, when properly planned and performed, validation helps to provide assurance that given QMS decisions and actions are effective. This chapter will explore the principles employed to develop and implement a risk-based Quality Management System that can be effectively linked to and supported by modern validation approaches.

"Quality is the result of a carefully constructed cultural environment. It has to be the fabric of the organization, not part of the fabric"—Phil Crosby (Duffy and Westcott, 2014). A QMS is not only a regulatory requirement within the Pharmaceutical Industry, but also paramount to the delivery of a reliable and quality manufactured product to the patient. A QMS should not be designed with the mindset of meeting a Regulatory Authority requirement but rather with the mindset of building a useful, efficient process that helps to determine the needs of the customer and transfers these needs into requirements that build a quality product. With this mindset, Quality

Principles of Parenteral Solution Validation. DOI: https://doi.org/10.1016/B978-0-12-809412-9.00004-6

truly does become a "fabric of the organization," not an additional process or "part of the fabric," intended to meet a requirement. With the end in mind, let's discover what an effective and efficient QMS looks like.

Per ICH Q10 (Pharmaceutical QMS), implementation of a QMS throughout the product life cycle should facilitate innovation and continual improvement and strengthen the link between pharmaceutical development and manufacturing activities (Eudralex, 2013). You might be asking yourself, "Is there such a thing as a QMS that facilitates innovation and continual improvement?" This does not seem possible when many in the industry are managing large quantities of deviations that are open for many days, vast numbers of change control activities that can take up to months to execute and complete, and Corrective and Preventive Actions (CAPAs) that are not being executed within the planned timelines. These are all symptoms that are a result of inefficiencies driven into the QMS. So how do we develop an efficient QMS that truly does drive innovation and continual improvement? The answer is using ICH Q9 (Quality Risk Management) in a pragmatic and meaningful way. By combining the power of ICH Q9 and ICH Q10, a QMS can emerge that is designed to drive continuous improvement initiatives, meet compliance requirements, and is efficient.

What benefits can be realized when ICH Q9 and ICH Q10 combine forces?

- Events are categorized based on the level of risk introduced to the product or process which allows flexibility in moving resources to manage the higher risk events.
- Project execution timelines and budget are based on prioritizing the mitigations of higher risk events.
- Risk-based self-inspection scheduling that ensures a frequency that meets the level of risk in an area.
- A balance between a proactive and reactive culture.

It is time for the industry to say goodbye to the current inefficient QMS processes often implemented. In the following sections, we will explore the integration of quality risk management (QRM) principles into a few focused areas of the QMS (see Fig. 5.1).

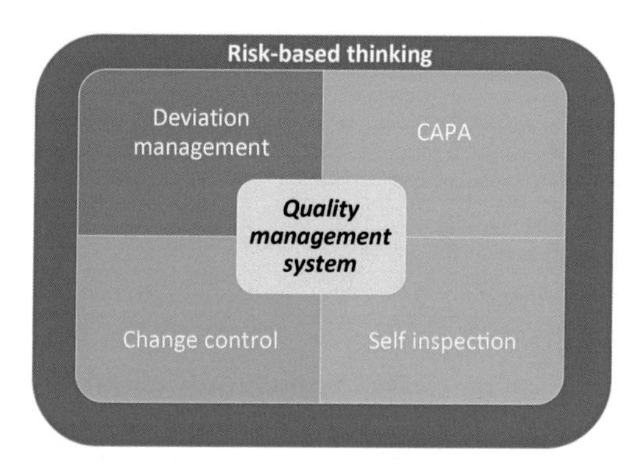

Figure 5.1 Quality risk management (QMS) focus.

Deviation Management

Deviations to validate processes should be evaluated as to their risk to the performance of the process and quality of the product. Where deviations require corrective or preventive actions, those actions may require additional validation studies to demonstrate their effectiveness and affect on the process performance and product quality. The level and extent of risk can be used to help determine and design effective validation studies. A Deviation Management System should include processes to manage feedback on product quality from both internal and external sources, (such as product complaints, product rejections, nonconformances, recalls, and deviations from good manufacturing practices (GMP) practices). The nature of a Deviation Management System is that it is reactive to events that have already occurred. These events may range in severity from product impacting, to minor departures from an approved procedure. Without a process to truly delineate the higher risk deviations from the lower risk deviations the resources will prioritize the workload on a First in First Out (FIFO) approach. FIFO may allow for high impact deviations to sit in a queue waiting for further assessment, while the less impactful deviations ahead of it, are being assessed with a Root Cause Analysis and plans for corrective actions. With a staff distracted and focused on less important items, high impact items will find their way into the QMS and may cause very serious implications for the firm.

Although QRM is best utilized in a proactive manner to determine control strategies to minimize the possibility of risk being realized, in the case of deviation management, it becomes issue management with a risk-based prioritization. Therefore, the best use of QRM methods is

Table 5.1 Example Deviation Categories

Risk Category	Definition
High	Event directly impacts critical quality attributes (CQAs), Critical Process Parameters (CPPs), compliance requirements, or will result in product quality impact
Medium	Event has the potential to indirectly impact CQA, CPP, compliance requirements or may result in product quality impact
Low	There is no impact to CQAs, CPPs, compliance requirements or product quality

to determine the rigor at which a deviation should be assessed. This approach will appropriately allocate resources and money to those deviation events which require a higher level of rigor. With the team focused appropriately, time can be dedicated to higher risk items to appropriately determine root cause and corrective actions.

The simplest approach for determining the actions needed to manage a deviation is to define a high, medium, and low risk definition (see Table 5.1 for example categories). Once these categories have been developed, they can be applied to the deviation, real time, working together between a Manufacturing Operations and Quality Assurance (QA) organization, as the event occurs.

The Risk Category is a quick and simple method that Operations working with QA, can perform on the shop floor. This approach can help determine decisions with respect to processing, forwarding, or halting the operations. It also allows for a quick triage process, so that appropriate actions can be taken immediately to address the event.

Low Risk Deviation

If, for example, a deviation event occurs, but is classified as Low Risk by QA, there is no need to halt operations or to seek further assessment. The deviation event is documented within the batch record, the QA person can review and approve, and operations commence. See Diagram 5.1 for an example decision tree in determining next steps, once a deviation event has been categorized. A Low Risk deviation batch comment should include a description of the event, the risk category, and rationale. Any additional steps taken to remediate the deviation event should also be captured.

Medium Risk Deviation

If a deviation event has been categorized as a Medium Risk, this event must be documented as a Deviation. The Deviation must include a

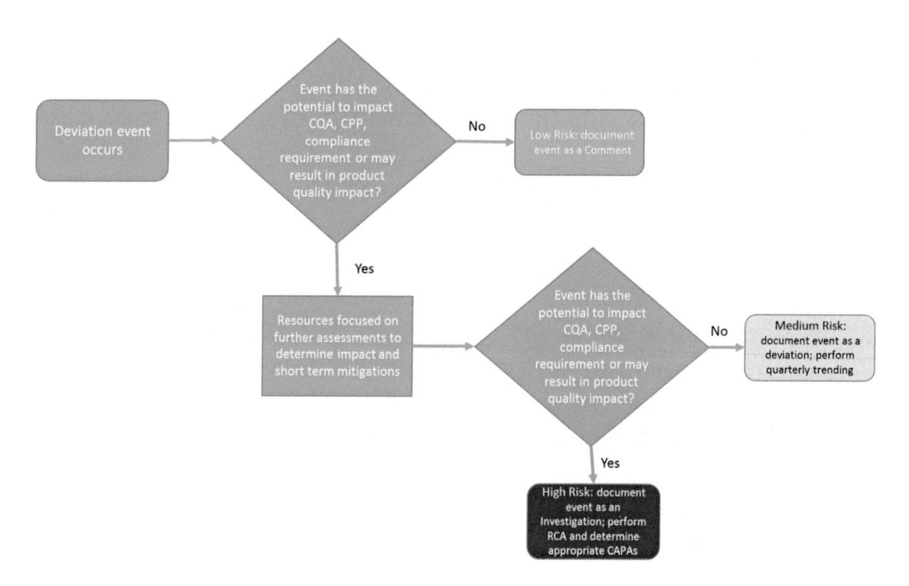

Diagram 5.1 Deviation management decision tree.

description of the event, the risk category, and rationale, and a summary of the assessments that were collected to determine the risk category. Any additional steps or actions taken to remediate the deviation event should also be captured. On a quarterly frequency, perform a Pareto analysis of all Medium Risk Deviations that occurred in the previous quarter. Trending should be performed to determine if events have occurred at a more frequent rate or if there is a possibility that the risk category will move to High (see Diagram 5.1). To best perform this analysis, a firm should have predetermined event categories that a Manufacturing Operator and/or QA can choose during the initiation of the Deviation. These categories can then be leveraged during the Pareto analysis. The top categories determined during the Pareto analysis should be assessed for root cause and corrective actions. The top Medium Risk Deviations are the potential signals that a High Deviation event may be realized in the future. A prospective management of the Medium Risk Deviations will minimize the realization of previously identified deviation events from becoming an out of control situation.

High Risk Deviation

A High Risk Deviation as per Table 5.1 has the potential to impact Critical Quality Attributes (CQAs), Critical Process Parameters

(CPPs), a compliance requirement or may result in product quality impact. To determine if a deviation event is high risk, the operator and QA may need to seek additional assessment from technical experts. Or it may be apparent during the operation that a high risk event has been realized. Either way, a decision must be made to determine if operations continue or cease, and/or if previous and current batches are impacted and their release decision. These discussions must occur in a timely manner. In addition to decisions that need to be made in the interim, a full investigation must occur, that requires a comprehensive root cause analysis, with CAPAs and CAPA effectiveness monitoring. If this is a repeat event, a failure analysis should be performed to determine why the previous CAPA implemented did not control the deviation event.

Corrective and Preventive Action

The CAPA is a system for implementing corrective actions for events that have been realized which may have resulted from events such as deviations, product complaints, or Health Authority observations. Preventive actions are actions that may prevent a negative impact event from being realized. Through the analysis of process monitoring data or prospective risk assessment, preventive actions are identified that may improve the process or reduce the possibility of a risk from being realized. When identifying CAPAs, it can become challenging to narrow down the best options and then to prioritize implementation. Whether you are a large manufacturing facility or in small volume clinical operations, there are many priorities to manage and a tool may be needed to prioritize CAPA efforts. Corrective and preventive actions may be taken as a response to a process validation study failure or a process failure. These actions represent to an extent changes to the process that would have to be incorporated in a re-qualification or re-validation response. The effectiveness of the CAPA at addressing the underlining issue, as well as an unintended consequences of the action should be taken into consideration. A risk based evaluation of the CAP, its effectiveness, and any effects on the process are therefore of importance.

Risk Benefit Analysis

A Risk Benefit Analysis, if performed correctly on all CAPAs, will provide valuable guidance in decision making.

Table 5.2 Risk Benefit Scoring Definitions

Risk Scoring Criteria	Severity Score	Opportunity Scoring Criteria	Benefit Score
May introduce potential product quality impact, Health Authority observation, or batch rejection	High	Eliminates Quality issue and risk outlined in Corrective and Preventive Action (CAPA), increases compliance reputation with Health Authorities	High
May introduce deviations with no potential for product quality impact, Health Authority recommendations	Medium	Moderately addresses Quality issue and risk outlined in CAPA, decrease in deviations, moderate increase in compliance reputation with Health Authorities	Medium
May introduce batch comments/ observations resulting from minor departures from GMPs,	Low	Partially addresses Quality issue and risk outlined in CAPA, very minor decrease in deviations, no impact to compliance reputation with Health Authorities	Low

Table 5.3 Example Risk Benefit Analysis

Option	Risk	Opportunity	Expense	Resource Hours (h)	Project Duration (months)
Option #1	High	Medium	$1Mm	500	5
Option #2	Medium	High	$500K	750	7.5
Option #3	Low	Low	$250K	600	4

- If several options are being evaluated to best address the CAPA intent, a Risk Benefit Analysis brings a standardized approach to evaluate all options equally. A decision can then be made by the team to determine the best option.
- The Risk Benefit Analysis provides the rationale and justification needed to present options to decision makers.

A Risk Benefit Analysis consists of two scoring criteria. One is the risk scoring criteria and the other is the benefit scoring criteria (Table 5.2).

The options that are being evaluated for the CAPA should be scored using both the risk and opportunity scoring criteria. In addition, expenses, resource hours needed, and duration of project should be assessed. With these criteria, a risk-based decision-making framework is established as well as a prioritization using the final score. See an example of the tool application in Table 5.3. In this example, the firm is evaluating three options for a potential CAPA. In these options, the risk, opportunity, project expense, resource hours, and project duration are assessed to determine which of the three options is best to mitigate the issue being assessed for a CAPA.

There is a lot of information to analyze for the decision-making process during this assessment. The three options contain information that can be broken down in the following summaries:

1. Option #1 seems challenging from the initial risk score. However the opportunity score is a Medium and sometimes with great risk can come great reward. In addition, the project can be completed with a minimal amount of resources and in less time. Yet the expense is significantly higher at double the next nearest option.
2. Option #2 is moderately risky with a Medium risk score, but rebounds with an opportunity score of High. The expense is half of the expense in Option #1, but the resource hours and project duration are significantly higher.
3. Option #3 is positioned the best with respect to risk, with a risk score of Low. However, the opportunity score was also Low, which brings to question whether this option will address our original quality issue well enough. The cost, resource estimates, and project duration are very low which is appealing from a business perspective.

With this information, what is the best decision? It depends on a few factors that a firm may be facing with respect to when the decision is made, and the risk tolerance of the decision makers. Remember making change always comes with a level of risk, it just depends on how much reward we may get with the risk that is taken. It also depends on how much risk we are willing to take. The power of the Risk Benefit Analysis used in CAPA delivers that information in a quick and easy tool. Option #2 with the Medium risk and High opportunity provides what the decision makers were looking for. While the resources and cost were higher than Option #3, the benefit made the option worth the wait and the firm may have needed a boost in compliance reputation. It is recommended with options identified to have Medium or High risk, to perform a more detailed risk assessment using a formal tool such as preliminary hazard analysis (PHA) or failure modes and effects analysis (FMEA). It is a best practice in general to perform a more detailed risk assessment prior to any project execution to account for not only Quality risks but other project-related risks that may pose a risk to project completion within the time and budget allotted. See Table 5.4 for the rationale to the decision made.

Table 5.4 Example Risk Benefit Analysis Decision Rationale

Option	Risk	Opportunity	Expense	Resource Hours (h)	Project Duration (months)	Decision/Rationale
Option #1	High	Medium	$1Mm	500	5	Choice #2—Risk is high and the benefit is medium. High cost
Option #2	Medium	High	$500K	750	7.5	Choice #1—Risk is medium, but the benefit is high. Cost is manageable
Option #3	Low	Low	$250K	600	4	Choice #3—Risk and cost may be low, but the benefit is low too

Developing a Risk Benefit Analysis tool can help to minimize bias in decision making and showcase the important elements needed to assess the best option. Table 5.4 becomes an Executive Summary that can then be shared with decision makers for discussion and final approval.

Change Control

Changes to systems and processes are an evitable part of doing business. Changes to the process may effect the output of the process and should be evaluated based on risk to process performance and product quality. Additional validation studies will likely be needed to include changes to the process.Risk assessment can be used to identify validation studies needed to evaluate the impact and effectiveness of changes. We are at times in a cycle of change to address an issue that needs to be fixed, but ideally, we are in a state of change due to continuous improvement. Whether we are improving the performance of our process or equipment, or introducing new changes, we must be in a state of control during the change process. To manage the volume at which change sometimes comes at us, it is ideal to implement a risk-based approach to change control oversight. Much like deviation management, if we are managing a large amount of changes with the same level of rigor, we may not be able to give the attention needed to evaluate each change. And we should not. Some changes are complex and require many project components to execute. However some changes are very minor and should not consume the time of expert resources to evaluate with in depth tools.

The first step to a simple risk-based approach to assessing change is by categorizing the change with a risk-based definition. Much like we

Table 5.5 Example Change Control Categories and Actions

Risk Category	Definition	Action
High	The change has the potential to adversely impact critical quality attributes (CQAs), Critical Process Parameters (CPPs), compliance requirement or may result in product quality impact	Additional risk management activities are *required* to assess the risks associated with the change
Medium	The change is being implemented to provide additional risk control measures to reduce the risk of impacting CQAs, CPPs, compliance requirements or product quality	Additional risk management activities are *recommended* to assess the risks associated with the change
Low	The change has no impact to CQAs, CPPs, compliance requirement or product quality	No additional risk management activities are required.

discussed how this can prioritize money and resources in a Deviation Management System, the same is true with Change Control. Similar scoring definitions from Deviation Management can also be applied for Change Control (see Table 5.5 for example categories).

The Risk Category is a quick and simple method that can help QA and technical Change Owners determine the level of rigor to assess the change pre- and postimplementation. If a change is Low risk, then follow the Quality procedures for implementing the change and no additional activities are needed. If a change is Medium risk, it is recommended to perform a formal risk assessment with a Risk Assessment Team to determine what additional risk control recommendations or decisions, need to be made prior to the change being implemented. If a change is High risk, this additional level of assessment is required. Tools such as PHA or FMEA are semiquantitative tools which provide the construct to risk rank and determine preventive and detection controls.

Change Control and Corrective and Preventive Action Effectiveness Monitoring

Once you have a final Risk Priority Number (RPN) from a semiquantitative tool such as PHA or FMEA, this number can also be used as the baseline value prior to implementing the change. The Risk Assessment Team should then determine the estimated postimplementation RPN value, based on the severity, probability of occurrence, and detection scores once the risk control recommendation has been implemented. The baseline preimplementation value and the estimated postimplementation value can be used to determine the effectiveness monitoring plan for the change. If the postimplementation RPN value is achieved then the change was effective. If the postimplementation RPN value is not achieved, then the change was not effective. This should trigger a review of the risk assessment and additional risk control recommendation conversations.

Once it has been determined that the change has been effective, the documentation can be updated to reflect the new risk assessment information and the change can be closed out.

Self-Inspection Process

As per EU Guidelines to Good Manufacturing Practice Medicinal Products for Human and Veterinary Use, Chapter 10, Preuse/Poststerilization Integrity Testing of Sterilizing Grade Filter (Self-Inspection), self-inspections should be conducted to monitor the implementation and compliance with GMP principles and to propose necessary corrective measures. In basic language, the self-inspection program is key to improving your QMS (EU, Brussels, 2010; Group, ICH Expert Working, 2008). Self-inspections are the opportunity to approach your QMS in a prospective manner and make improvements that drive compliance efficiency into the program. As per Chapter 10, Preuse/Poststerilization Integrity Testing of Sterilizing Grade Filter, a self-inspection program should cover the following:

- Personnel matters, premises, equipment, documentation, production, quality control, distribution of the medicinal products, arrangements for dealing with complaints and recalls, and self-inspection, should be examined at intervals following a prearranged program to verify their conformity with the principles of QA.
- Self-inspections should be conducted in an independent and detailed way by designated competent person(s) from the company. Independent audits by external experts may also be useful.
- All self-inspections should be recorded. Reports should contain all the observations made during the inspections and, where applicable, proposals for corrective measures. Statements on the actions subsequently taken should also be recorded.

A self-inspection team, which should be trained and highly skilled in auditing principles, should follow a checklist approach in preparing for the audit. The team should follow an annual schedule that ensures

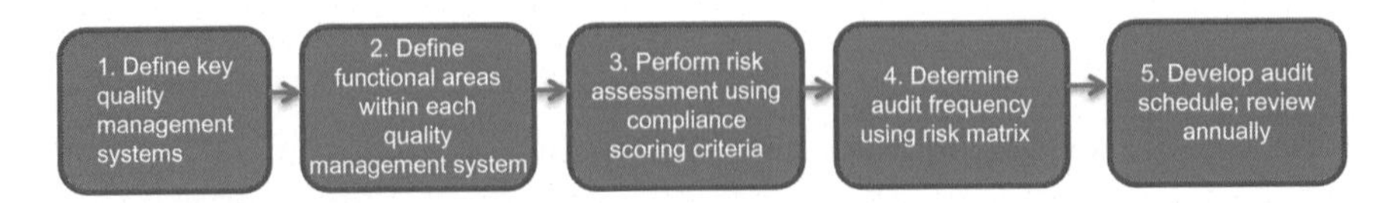

Diagram 5.2 Risk-based self-inspection process.

the applicable areas are audited at an appropriate frequency, supported by a justification/rationale. In addition, the processes within a specific area should be highlighted and it should be documented why these processes were chosen to be within the scope of the self-inspection. If risk management is used appropriately to determine the audit schedule and frequency, benefits will emerge from performing self-inspections and determining continuous improvement initiatives within the program.

To establish a risk-based self-inspection program, the following process in Diagram 5.2. *Risk-Based Self-Inspection Process* must be completed for the entire site or function to establish the initial plan. Once the plan is established, it must be reviewed on an annual frequency to identify any changes made that may impact the outcome of the risk assessment performed. This in turn, may result in a change to the plan and/or audit frequency.

Define Key Quality Management Systems

The first step is to define the Key QMS within your site, organization, or functional group. The list below is an example of how this can be performed. This is an example of how to align key QMSs but may be modified based on the business operations or organizational design. The purpose of defining these systems is to break a large facility into manageable pieces that you can then break into further subcategories in the next step to build a realistic scope for your self-inspection program.

- Production;
- Facilities, Equipment, Utilities;
- Laboratory Control and Operations;
- Materials;
- Quality; and
- Packaging and Labeling.

Table 5.6 Focus Areas by Quality Management System

Quality Management System	Focus Area(s)
Production	Upstream Manufacturing Process
	Downstream Manufacturing Process
	Technology Support
	Process Development
	Engineering
	IT
Facilities, Utilities, and Equipment	Maintenance Operations
	Instrument Services/Metrology
	Production Planning and Inventory Management
	Utilities Operations
	Validation
	Buildings and Grounds (includes Pest Management)
Laboratory Control and Operations	QC Testing—Product
	QC Testing—Environmental
	QC Testing—Raw Materials
Materials	Raw Material Receipt, Storage and Distribution
	Product Storage and Distribution
Quality	CAPA
	Change Control
	Deviation Management
	Training Program or Records
	Batch Release
	Batch Record Review

Define Functional Area Categories

To appropriately cover all processes within the QMS, subcategorize the various functions performed within each of the Key QMSs listed above (Table 5.6). This exercise will assist in building a self-inspection program that is manageable in size and scope. For example, if you are performing a self-inspection on a biologics Active Pharmaceutical Ingredient facility with 1000 employees and an annual production schedule that contains 5 high volume drugs, the facility is extremely busy and will not desire to perform inefficient, lengthy audit activities that are not appropriately focused. By breaking the Production system into Upstream, Downstream, Engineering, Technology Support, IT, and Process Development, the audit activities will have a right-sized focus and will be able to spread these functional areas out over the next few years so that the site stays focused on the highest risk areas first and more frequently, within Production. Without this delineation, the audit may lose focus and will not be adequate to thoroughly evaluate a focus area.

Perform Risk Assessment Using Compliance Scoring Criteria

Now that the QMS categories have been developed and the Focus Areas have been linked to each QMS, perform the risk assessment using the Compliance Scoring Criteria. It is important to note that there are many ways you can build a risk-based self-inspection audit program. The definitions outlined in Table 5.7. *Risk Assessment Criteria and Definitions* is an example of how this can be defined. The Risk Factors themselves can also vary in category definition as well. Below is an example of how these three categories were defined and why they were chosen for this risk assessment activity.

Risk Factors and Scoring

- *Compliance Status*—provides an opportunity to assess a focus area based on previous audit history performance. This category allows an evaluation over time to see if the compliance status of an area changes. If the compliance status score becomes low, then the self-inspection frequency should lower as well. The inverse would be true for a high compliance score as well.
- *Criticality to Production*—places the functional area in proximity to the patient. The assumption being, the closer the operation is to the final distribution of a drug product, the more risk it potentially imparts on the patient. This category allows some risk ranking based on the operation or functional area itself and the inherent risk it imposes, regardless of the compliance status or audit history.
- *Audit History*—gives a snapshot review of the audit history of a functional area. The less of an audit history, the more uncertainty with proving its defense, which may translate to risk. This category provides an opportunity to evaluate not just the compliance status but how frequently an area has been assessed as well.

Assess each focus area within the QMS for the Compliance Score, Criticality Score, and Audit History Score. Multiply the scores together to calculate the overall Risk Level (see example in Table 5.8).

Table 5.7 Risk Assessment Criteria and Definitions

Compliance Status		Criticality to Production		Audit History	
Criteria	**Rank**	**Criteria**	**Rank**	**Criteria**	**Rank**
Critical Compliance Risk • Repeat inspection/audit observations or CAPAs not completed per commitments • Previous identification of Critical Audit Observations • Currently subject to Regulatory or market action	10	Very Critical • Directly incorporated or responsible for production of drug product	10	No Previous Audit History/Latest audit > 3 years • Has no previous audit history on record either from Health Authority, internal audits or customers? • Has previous audit history on record; however, it has been greater than 3 years since last audit	10
Major Compliance Risk • Quality Management System elements exist, but are not completely present, consistently or effectively applied as highlighted by Audit findings • Previous identification of Major Audit Observations • Deviations from dossier filing or GMPs • Investigation events have occurred in the area	8	High Criticality • Directly incorporated or responsible for production of Active Pharmaceutical Ingredient (API) • Final release testing of API or critical raw materials	8	Not Audited Within Previous 2 Years • Has not had an internal audit in the past, but has had Health Authority inspections or customer audits within 2 years • Has been internally audited in the past, but not within 2 years	8
Minor Compliance Risk • Quality Management Systems exists, but inconsistent cross-site practices exist • Previous identification of Minor Audit Observations per QA/SY/2.40 • Minor departure from GMPs or inadequate procedures as highlighted by audit findings	6	Critical • Directly supporting production of API • Directly supports Environmental conditions during operations	6	Not Audited (or not currently planned) within Current Year • Has not been internally audited in the past, but has had Health Authority inspection within 1 year • Has been internally audited in the past, but not within the last annual audit period	6
Low Compliance Risk • An isolated minor departure from GMPs or inadequate procedures may exist • Recommendation(s) from previous audit(s)	4	Low Criticality • Indirectly supports production of API • In process control testing of API or support operations necessary for lot release or operations	4	Quality Management System Audited (or currently planned) within previous or current period, Not Specific Function • Larger Quality Management System was internally audited, but the specific function was not covered by an internal audit • Specific Functional area was inspected by a Health Authority within calendar year • Internal audit last performed within 2 years	4
No Compliance Risk • Compliant with all policies, procedures and GMPs • Non-GMP operations	2	Noncritical • Supporting Facility or Utility based operations	2	Specific Function Audited (current period) • Internal audit has been performed within previous or current annual audit period	2

Table 5.8 Risk Assessment Template Example

Quality Management System	Focus Area (s)	Compliance Assessment	Compliance Score	Criticality Assessment	Criticality Score	Audit History Assessment	Audit History Score	Risk Level
Production	CAPA	Quality Management Systems exists, but inconsistent cross-site practices exist	6	Indirectly supports production of Active Pharmaceutical Ingredient	4	Has been internally audited in the past, but not within 2 years	8	192

Table 5.9 Determine Audit Priority and Frequency

Risk Level	Priority	Frequency
8–64	Low	Every 2 years
72–160	Medium	Annual
≥ 192	High	6 months

Table 5.10 Priority and Frequency for Audit Plan

Quality Management System	Functional Area (s)	Risk Level	Priority	Frequency	Year 2018	Year 2019	Year 2020
Production	CAPA	192	High	6 months	March	September	March

Determine Audit Frequency Using Risk Matrix

Once you have performed the risk assessment for each Focus Area, leverage the example Risk Matrix in Table 5.9 to determine the priority and frequency for each area.

Develop Audit Schedule; Review Annually

To develop the audit schedule, utilize the priority and frequency recommendations in Table 5.9 for a proposed 3-year schedule (Table 5.10). *Priority and Frequency for Audit Plan*, for an example of how to document the audit plan for a focus area. Once all focus areas have been listed with the priority and frequency, you will need to ensure resources are available to complete the audits.

Annually, revisit the Risk Assessment and Audit Plan to determine if changes need to be made. Events that trigger a review of the Audit Plan are as listed (not an all-inclusive list):

- Changes to the Risk Level;
- New or existing focus areas; and
- Organizational or department functional changes.

The purpose of the risk assessment is to determine the frequency at which a functional area should be considered for audit and to determine the prioritization of the audits for the year. This risk assessment should be performed and updated on an annual basis to ensure that as the landscape changes; the audit plan reflects the current state.

Summary

Someone once said, the definition of insanity is doing the same thing over and over again and expecting a different result. If your QMS is no longer positively driving the business and aiding in continuous improvement initiatives, or your organization is lost in a sea of deviations, change control records and overdue CAPAs, now is the time to do something different. By implementing QRM principles in the QMS, building a useful, efficient process that helps to determine the needs of the customer and transfers these needs into requirements that build a quality product, may finally become the focus of the QMS that it always should have been.

References

Duffy, G.L., Westcott, R.T., 2014. The Certified Quality Improvement Associate Handbook, third ed.

Eudralex, 2013. Volume 4, EU Guidelines for Good Manufacturing Practice for Human and Veterinary Use. Chapter 1, Pharmaceutical Quality System.

EU Guidelines to Good Manufacturing Practice Medicinal Products for Human and Veterinary Use, Chapter 9 (Self-Inspection), 2010. European Commission Health and Consumers Directorate-General, Public Health and Risk Assessment: Pharmaceuticals. Brussels, SANCO/C8/AM/sl/ares(2010)1064587.

FDA's Drug Manufacturing Inspection Compliance Program.

Group, ICH Expert Working, 2008. Pharmaceutical Quality System Q10.

Further Reading

ICH Q9, 2005. Quality Risk Management.

CHAPTER 6

Use of Statistics in Process Validation

Igor Gorsky
Senior Consultant, ConcordiaValsource LLC., Downingtown, PA, United States

Overview

When Food and Drug Administration CDER, CEBR, and CVM team issued revised guidance for industry on Process Validation (FDA Guidance for Industry, 2011) they "recommend an integrated team approach[1] to process validation that includes expertise from a variety of disciplines (e.g., process engineering, industrial pharmacy, analytical chemistry, microbiology, statistics, manufacturing, and quality assurance)." Interestingly that the discipline of "statistics" is listed prior to disciplines of "manufacturing" and "quality assurance." We believe that an order in which these disciplines are listed is important as it is once again reflects FDA's objective to convey Risk-Based Life Cycle approach for the industry. Therefore by this order it is important to establish once and for all, an important role statistic should play in parenteral pharmaceuticals design and development, qualification, and continued verification.

Statistics is the science of collecting, organizing, and interpreting numerical facts, which we call data. In today's world we are surrounded by information. Every year the number of servers on the World Wide Web grows exponentially. If in 2005 there were 70 million sites on Internet, in 2014 there were more than a billion. As one textbook on statistics says, "we are bombarded by data in our everyday life (Moore and McCabe, 2005)" and further adds that "a knowledge of statistics helps separate sense from nonsense in the flood of data." The fundamental goal and objective of statistics is to listen to and to understand data collating into information from which we may gain knowledge which may eventually make us wiser. To gain understanding, we often operate on a set of numbers—we average or graph them,

[1] This concept is discussed in more detail in Ref. FDA's Guidance for Industry.

Principles of Parenteral Solution Validation. DOI: https://doi.org/10.1016/B978-0-12-809412-9.00005-8

for example. But we must do more, because data are not just numbers; they are numbers that have some context that helps us understand them.

We can examine main reasons for use of statistics in parenteral pharmaceutical manufacturing and Process Validation since these subjects go hand-in-hand.

First and foremost, we need to realize and understand that our processes are variable and not set in stone. It is not a matter of will it, but a matter of when it will shift. Processes shift due to equipment getting older, hiring of new employees and not able to capture knowledge through effective training programs, variability of raw materials, variability of due to analytical instrumentation, and many other things. Sometimes shift occurs due to combination of different parameters and it is called multivariate variability. It does not mean that process is not under control. This only means that there is a difference from lot to lot and within lots. That is the difference that we are trying to understand. As the phrase attributed to Dr. W. Edwards Deming says—"Uncontrolled variation is the enemy of quality." Therefore we need to measure and control process variation. Statistics helps us measuring and understanding it.

Second reason for a use of statistics in parenteral manufacturing and Process Validation is to achieve and maintain consistency of the manufacturing processes. As FDA Guidance for Industry on Process Validation states "process validation is defined as the collection and evaluation of data, from the process design stage through commercial production, which establishes scientific evidence that a process is *capable of consistently delivering* quality product" (FDA Guidance for Industry, 2011). The only way to measure capability of consistent delivery of quality product is to understand variability, have ways to control it and continue monitoring it through a life cycle. Although the level of consistency may have a certain degree of variation, it should be predictable based on our knowledge and measurement of the process.

Third and the most important reason to embed statistical discipline in Process Validation program is to assure that new products use robust design. Reduction of the variation in the product/process involves not just setting it on target but trying to find conditions of

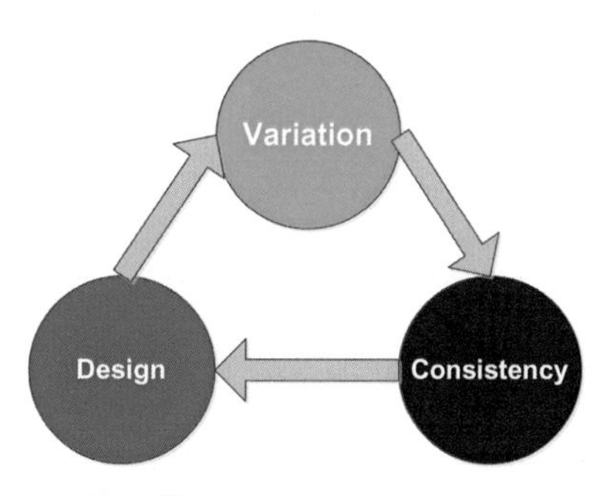

Figure 6.1 Statistics use objectives in Parenteral Process Validation.

design factors which make the product/process insensitive or robust to its environment. The idea of a robust designed was popularized by Taguchi (1995). He opened an entire field that previously had been only discussed theoretically with very few applications in real life. His quality engineering methodology relied on implementation of design of experiments which concentrate not just on the consistency of mean but also the variance.[2] It is a method that has been widely accepted by the regulators and described in an ICH Q8 Guidance (ICH, 2009a). In this guidance, ICH refers to this method as a Quality by Design which it defines as "a systematic approach to development that begins with predefined objectives and emphasizes product and process understanding and process control, based on sound science and quality risk management."

Of course, these major reasons for using statistics in Parenteral Process Validation could have been and should have been listed in a reverse order, as process design and development is the first stage of Process Validation life cycle concept; however, this author is listing them in this particular order was pursuing a goal of introducing statistical concepts that are required to be understood first prior to proceeding to next steps, such that variability needs to be understood first prior to proceeding to consistency and quality by design. These concepts should all constitute a life cycle approach and are illustrated in Fig. 6.1 as a continuum.

Concepts outlined above are pillars on which statistical quality control should be based; however, regulators also have specific requirements

[2] https://www.nist.gov/sites/default/files/documents/2017/05/09/robdesgn-1.pdf.

in guidances that should be followed and immediately addressed in firms' quality systems.

We will list only the most important of these requirements which are covered in FDA's Process Validation guidance. The most obvious reference to the utilization of statistical methods is mentioned in the beginning of the document which lists 21CFR211 CGMP requirements that discuss important aspects that must be included into validation program. Namely, Process Validation Guidance states "other CGMP regulations define the various aspects of validation. For example, § 211.110(a), *Sampling and testing of in-process materials and drug products*, requires that control procedures '... be established to monitor the output and *to validate* the performance of those manufacturing processes that may be responsible for causing variability in the characteristics of in-process material and the drug product' (emphasis added). Under this regulation, even well-designed processes must include in-process control procedures to assure final product quality. In addition, the CGMP regulations regarding sampling set forth a number of requirements for validation: samples must represent the batch under analysis [§ 211.160(b)(3)]; the sampling plan must result in statistical confidence [§ 211.165(c) and (d)]; and the batch must meet its predetermined specifications [§ 211.165(a)]." Interestingly a lot of firms overlook these requirements even now, 6 years after FDA Guidance publishing. One can find similar to these Form 483 Observations while reviewing current FDA inspections—"Failure to demonstrate that manufacturing processes can reproducibly manufacture drug substances meeting predetermined quality attributes. Process validation for a drug substance did not establish sound sampling plans to evaluate variability among batches. The report notes that this observation was also made in a March 2015 Form 483" (Kelly, 2017).

Another important aspect mentioned in the FDA's Guidance is use of statistics while establishing process and product specifications. It specifically states—"In addition to sampling requirements, the CGMP regulations also provide norms for establishing in-process specifications as an aspect of process validation. Section 211.110(b) establishes two principles to follow when establishing in-process specifications. The first principle is that '... in-process specifications for such characteristics [of in-process material and the drug product] shall be consistent with drug product final specifications ...' Accordingly, in-process material should be controlled to assure that the final drug product will meet its quality requirements. The second principle in this regulation further requires that in-process specifications '... shall be derived from previous acceptable process average and process variability estimates where possible and determined by the application of suitable statistical procedures where appropriate.' This requirement, in part, establishes the need for manufacturers to analyze process performance and control batch-to-batch variability." We will examine several analyses that could be used for this purpose.

In addition to Process Validation Guidance other guidances, such as Guidance for Aseptic Processing (FDA Guidance for Industry, 2004a) also mention statistics, but in a slightly different light, outlining shortcomings of USP Sterility method sampling. However, in order to supplement this USP sampling as recommended in this guidance one needs to understand statistical concepts used to derive sampling plans listed in Aseptic Processing Guidance. In addition to mentioned passages from regulatory guidances, there are a great number of others from which I would like to specifically mention two—"WHO guidelines for sampling of pharmaceutical products and related materials" (World Health Organization, 2005) which stipulates some statistical sampling concepts and GHTF Quality Management Systems—Process Validation Guidance for Medical Devices (GHTF/SG3/N99-10:2004, 2004) which has a great section on statistics. Although the last guidance is for medical devices it is definitely applicable for Parenteral Process Validation, as well.

In the next few pages of this chapter we would like to examine some of the fundamental applications of statistics in Process Validation. Since we answered a question, why do we need statistics? For one, it is a regulatory requirement, as was illustrated earlier. In addition, we can learn a lot from enumerating our results and analyzing them using statistical method which should help with our product knowledge and process understanding that in turn are a fundamental bases of Pharmaceutical Quality System (ICH, 2009b).

Next, we will try to answer a question—what do we do with statistics? Our answer will include a few statistical concepts which is not designed to give any in-depth analyses, but instead concentrate on a more simplistic principals, that support implementation of statistical discipline in Process Validation. We think that an important part of listing and description of these principals is presented by an author who spend most of his long career observing, reviewing, analyzing, and investigating processes and developed a passion for a "story" each one of them is ready "to tell" to those who are willing to "listen."

Use of Statistics in Parenteral Process Validation—Ten Basic Concepts

First Concept—A Concept of Statistical Inference

One should understand that, as discussed earlier, sampling should be statistically significant and should be representing product or process population. However one should also understand a Concept of Statistical Inference which a theory, a method, and a practice is of making judgments about the attributes and parameters of a product or a process population and reliability of statistical relationships between the parameters and attributes (e.g.), which is typically based on random or statistically significant and representative sampling. In other words, one bases its decision for commercial release of the product, for instance, on a representative sampling and typically and routinely does not really sample 100% of the product unless one is using Process Analytical Technology. It is an important concept to understand because sampling is only a representation and because of the inherent variability of the product, process, and materials constituting the product only gives an estimate of a true means and a true variance, as well as statistical relationships.

Second Concept—Graph It First!

When one starts on a subject of implementation of statistical concepts one should first and foremost learn how to graph data. The graphing of data transforms it from mere numbers into an often a colorful picture which has been known for centuries to "worth a ten thousand words." We are not going to go into details of what kind of graphs could be used for which purpose, but rather we will leave it for practitioners implementing statistical methods. However a few graphs that could help summarization of your data could be briefly mentioned, namely boxplots (representing a statistically plotted data based on the minimum, first quartile, median, third quartile, and maximum of population that looks like a box and allows one to examine data distribution), individual value plot (shows a dot representing an actual value of each observation in a population which also makes distribution spread very obvious), and a scattered value plot (which shows the values of two variables that are plotted along *x*- and *y*-axes and quickly visually reveals if a pattern resembles correlation between these variables).

Obviously, there are many more plots that could be used to illustrate and visually represent your data. We recommend trying to use different charts to characterize your data as they may visually show different aspects of studied populations and some could be more useful than the others. The goal of this brief review is to bring attention to this valuable statistical tool.

Third Concept—Let DataTalk to You!

Upon plotting the data and examining plots one may observe accuracy and precision of "hitting" one's target, see if there are any outliers in data sets, and visually observe studied populations. This may allow one to "let data talk to you." Data if analyzed properly always have a story to tell. Data from different lots may look the same or it may look different, but it always tells the story of the level of variability, consistency and robustness of design of the product and the process. One's responsibility is to "listen" to data and use it to transform into information which in turn builds product knowledge and leads to process understanding (Fig. 6.2).

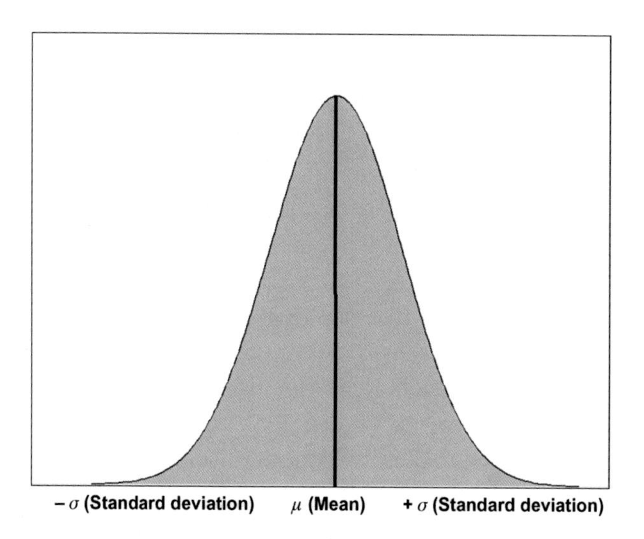

Figure 6.2 Normal distribution illustration.

Fourth Concept—Normality?

When one talks about normality of data one describes a test that defines how normal product and/or process data population is distributed. When data are plotted using a histogram plot, for instance, a normal distribution will look like a bell-shaped curve both sides of which will be symmetrical about its mean. It is also assumed that the normal distribution is the most common statistical distribution "because approximate normality arises naturally in many physical, biological, and social measurement situations."[3] Many statistical analyses are typically insisting on data being normally distributed. The normal distribution is also called Gaussian distribution, named after great German mathematician who first described method of least squares and method of maximum likelihood which are the bases for normal distribution concept. In short, normal distribution is defined by the mean and the standard deviation, where mean is "the peak or center of the bell-shaped curve and the standard deviation determines the spread in the data."[4]

One of the main reasons of mentioning normality of population distribution here was not just constatation of the fact that it is one of the main concepts of statistical science, but also to turn readers attention to those instances when one encounters a nonnormal distribution. It should be noted that the subject of nonnormality comes to the attention when one uses data distribution for statistical analysis such as process capability indices (PCIs) calculation. Therefore we would like to quote here one of the definitive books on PCIs that examines issue of distribution normality and comes to the conclusion that normality may not be of a primary importance. This volume states—"The discussion of nonnormality falls into two main pans. The first, and easier of the two, is investigation of the properties of PCIs and their estimators when the distribution of X has specific nonnormal forms. The second, and more difficult, is development of methods of allowing for nonnormality and consideration of use of new PCIs specially designed to be robust (i.e., not too sensitive) to nonnormality. McCoy (1991) regards the manner in which PCIs are used, rather than the effect of nonnormality upon them, as of primary importance and writes: 'All that is necessary are statistically generated control limits to separate

[3] Minitab Statistical Software, ver. 17, Help "Normal Distribution" article.
[4] Minitab Statistical Software, ver. 17, Help "Normal Distribution" article.

residual noise from a statistical signal indicating something unexpected or undesirable is occurring.'"(Kotz and Johnson, 1993) Furthermore it could be recommended, most importantly, to analyze data distributions, even if they are nonnormally distributed, continually in time using the same statistical methods every time to evaluate changes and determine trends.

Fifth Concept—Descriptive Statistics

Another concept that is always useful to utilize in the beginning of data analysis is descriptive statistics to represent one's data. The descriptive statistics are a number of coefficients that summarize any obtained data set. These coefficients could be representing either an entire data population or a sample of that population from which one would build her inference of that population. Although descriptive statistics could be performed on univariate, bivariate, or multivariate analyses, we will briefly state those statistics that are typically calculated and presented in only for a univariate analysis. These statistics are typically measuring several coefficients that listed below:

- mean, median, and mode, that are describing the central tendency which is a typical representative of a population probability distribution;
- range and quartiles of the sampled data set representing a dispersion of the population;
- variance, standard deviation, and RSD which measures a spread of the population.

There are many other statistics that describe sample sets and product/process studied populations; however, our goal is only to bring attention to this useful concept.

Sixth Concept—Control Charting

Control charts were created by Dr. Shewhart in early 1920s at Bell Telephone Laboratories (part of Western Electric) which, by the way had what we now call a mission that stated, "as alike as two phones." Obviously phone manufacturing company was striving for consistency. A noble goal which is about 100 years later we (in parenteral manufacturing) should be striving for as well. Dr. Shewhart uncovered

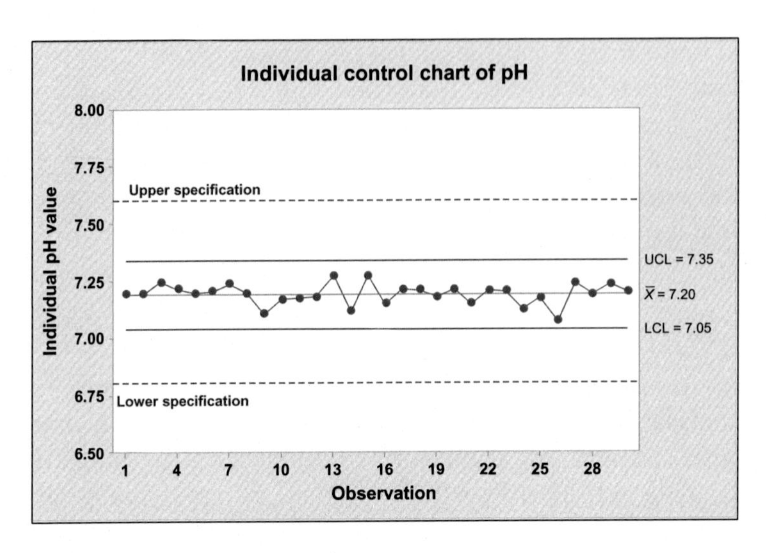

Figure 6.3 Statistical process control chart example.

a principal on which understanding of the process variation could be based and answered by control charting. The principal was based on understanding of causes of variation which are two. First causes of process variation are called common causes. If process parameters and/or product attributes are plotted on control chart these result points will always be within controlled process. They are systemic inherent process variation. These are the causes that we are looking for, for instance, when we are analyzing legacy processes (Fig. 6.3).

These causes we want to understand when we are trying to optimize the processes. The second causes of process variability are special causes. These causes are not systemic. They make one's result points plotted on the control charts go out of control limits.

These are the causes that are harder to understand, thus harder to predict. Dr. W. Edwards Deming in his book *The New Economics for Industry, Government, Education* goes further in analyzing two common mistakes regarding control charting.

1. Mistake 1: To react to an outcome as if it came from a special cause, when it came from common causes of variation.
2. Mistake 2: To treat an outcome as if it came from common causes of variation, when it came from a special cause.

We find it interesting that these mistakes are very often made by parenteral drug manufacturers in process validation of the parental products, as well as in commercial production. Manufacturers tend to overreact after they are being cited by regulators and typically make mistake 2, while during regular commercial production they often make mistake 1. Many firms tend not to subscribe to process control reviews finding them unnecessary burden which may increase costs of operations. Those companies are typically paying high costs in batch rejections, regulatory citing, as well as loss of knowledge understanding. Therefore, we would highly recommend introducing this statistical concept as early in the process validation life cycle as possible. In addition, it is important to mention careful consideration and use of statistical process control rules such as Western Electric and Nelson's rules to make review of your SPC meaningful.

Seventh and Eight Concepts—Tolerance and Capability

We will discuss seventh and eight concepts in one section illustrating their usefulness, ability to detect process variability early in the process validation life cycle and relationship to each other. Because these concepts are related it is important to utilize them in tandem. We will be using two examples that show how usage of these concepts early on for setting proper specifications and ability of the process to be consistent or inconsistent based on the set specifications. First, we will define tolerance intervals. Tolerance interval is an "interval determined from a random sample in such a way that one may have a specified level of confidence that the interval covers at least a specified proportion of the sampled population."[5] These intervals should be helpful when we are analyzing process design data and developing specifications. Using this statistical concept complies with FDA's Process Validation Guidance which states that specifications "shall be derived from previous acceptable process average and process variability estimates where possible and determined by the application of suitable statistical procedures where appropriate" (FDA Guidance for Industry, 2011).

The general tolerance interval equation that are listed in International Standard ISO 16269 are shown in Eq. (6.1) below:

[5] International Standard ISO 16269-6 Second edition 2014-01-15 Reference number ISO 16269-6:2014(E) Statistical interpretation of data—Part 6: Determination of statistical tolerance intervals statistical tolerance interval, Brussels, 2014.

$$y = \overline{X} \pm k \times S \tag{6.1}$$

Equation 6.1 Tolerance Interval.

where y is a tolerance interval; $\overline{X}$ is a sample mean $\overline{X} = 1/n \sum_{j=1}^{n} x_j$; k is a factor used to determine the limits; S is a sample standard deviation $S = \sqrt{1/(n-1) \sum_{j=1}^{n} (x_j - \overline{x})^2}$

Next, we will use above equation to calculate a tolerance interval for an example of Bottle Torque results for a Large Volume Parenteral product attained during packaging process which is a Critical Quality Attribute (CQA) for a product. Although, bottle cap torque is not a typically useful attribute for parenteral products; nevertheless, we will use this example for ease of illustration. Table 6.1 shows results of torque testing during packaging process design phase (stage 1 of Process Validation life cycle). The hypothetical firm's QC department decides to set a specification range for this CQA at 5–12 kp (kilopond) as none of the results were outside of these values. This firm "X"'s QC group did not perform any significant data analysis and only calculated mean and standard deviation of the sample set. Then they followed a thinking process that typically used $\overline{X} \pm 3\sigma$ for setting specifications would result in a too wide of a specification and that rounded to nearest whole number $\overline{X} \pm 2\sigma$ would be more appropriate. In addition, team was convinced that because all the attained results would be passing chosen limit it was appropriate. However if they were performing tolerance intervals calculations they would have easily found out that their chosen limit is statistically too tight and either they need to do a capping/re-torqueing machines adjustment (desired path) or they would have to choose a wider limit. The tolerance interval calculation for this example is shown in Table 6.1 and Fig. 6.4 and were using k-factor values for 10 samples and 15 samples for selected confidence interval of 95% and 95% and 99% of population.

Continuing with this hypothetical example, above firm "X" performed a Process Performance Qualification study in which a statistically significant number of samples (30) was taken for a cap torque tests to qualify a cap torqueing process and to understand process variability. The results of this study are shown in Table 6.2 and Fig. 6.5.

Table 6.1 Large Volume Parenteral Product Tolerance Example Summary

Product: ***Large Volume Parenteral Product*** **Data CQA:** ***Bottle Torque*** **Sample No.**	**Result (kp)**	
1	6.7	
2	9.5	
3	7.7	
4	6.7	
5	6.8	
6	11	
7	11.1	
8	6.2	
9	10.7	
10	8.2	
Determine mean (x-bar)	8.46	
Determine standard deviation (S)	1.95	
Specification minimum	5	
Specification maximum	12	
Confidence level; 95.0%		
	10 Samples	15 Samples
k-constant for 95% of Population estimate	3.394	2.965
Calculated minimum tolerance	1.84	2.68
Calculated maximum tolerance	15.08	14.24
	10 Samples	15 Samples
k-constant for 99% of Population estimate	4.437	3.886
Calculated minimum tolerance	0.00	0.882
Calculated maximum tolerance	17.11	16.04

It is important to note that FDA Guidance recommends that "production data should be collected to evaluate process stability and capability" (FDA Guidance for Industry, 2011). Process capability and process performance indexes, Cp and Pp are indexes that describe process capability or performance in relation to a specified tolerance (ASTM E2281).

In addition to evaluation of a process capability requirement, FDA Guidance lists ASTM E2281 "Standard Practice for Process and Measurement Capability Indices," and ASTM E2709 "Standard Practice for Demonstrating Capability to Comply with a Lot Acceptance Procedure." as useful references for this analysis. As shown

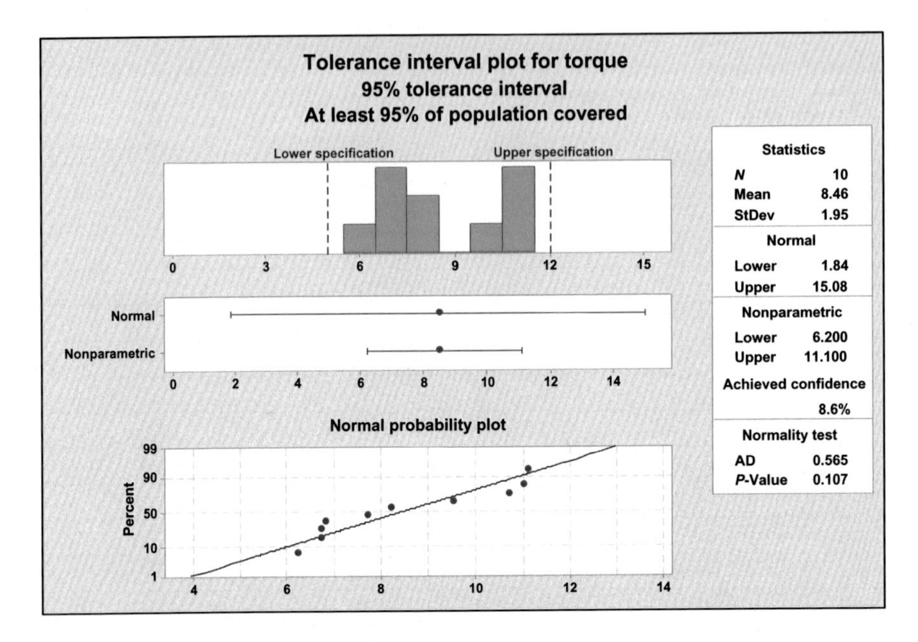

Figure 6.4 Large volume parenteral product tolerance example plot.

Table 6.2 Large Volume Parenteral Product Capability Example Summary

Product: *Large Volume Parenteral Product* Data CQA: *Bottle Torque* Cpu equation: Cpl equation: Cpk Sample No.	 *Cpu = (UCL − x-bar)l3 × S* *Cpl = (x-bar − LCL)l3 × S* *Min (Cpu, Cpl)* Result (lb/in.)
1	7.7
2	8.1
3	12.1
4	5.6
5	6.7
6	8.1
7	8.6
8	6.2
9	8.9
10	7.2
11	9.4
12	8.6
13	3.3

(Continued)

Table 6.2 (Continued)	
Product: ***Large Volume Parenteral Product*** **Data CQA:** ***Bottle Torque*** **Cpu equation:** **Cpl equation:** **Cpk** **Sample No.**	 ***Cpu = (UCL − x-bar)/3 × S*** ***Cpl = (x-bar − LCL)/3 × S*** ***Min (Cpu, Cpl)*** **Result (lb/in.)**
14	10.0
15	9.3
16	8.4
17	10.3
18	7.4
19	9.3
20	6.5
21	8.6
22	7.7
23	5.9
24	7.9
25	8.4
26	10.1
27	3.4
28	8.4
29	5.9
30	8.8
Determine mean (x-bar)	7.90
Determine standard deviation (S)	1.90
Specification minimum	5
Specification maximum	12
Confidence level; 95.0%	
Cpu	0.72
Cpl	0.51
Cpk	0.51

in several results (highlighted) were out specification and PCIs (on the plot presented as process performance Ppk) yielded only 0.51 (a desired 4σ process yields Cpk = 1.3).

In this simple example we have shown usefulness of tolerance interval and process capability concepts, as well as their relation to each other. When firms implement tolerance interval analysis early on in a design phase of the Process Validation, it allows them to assess

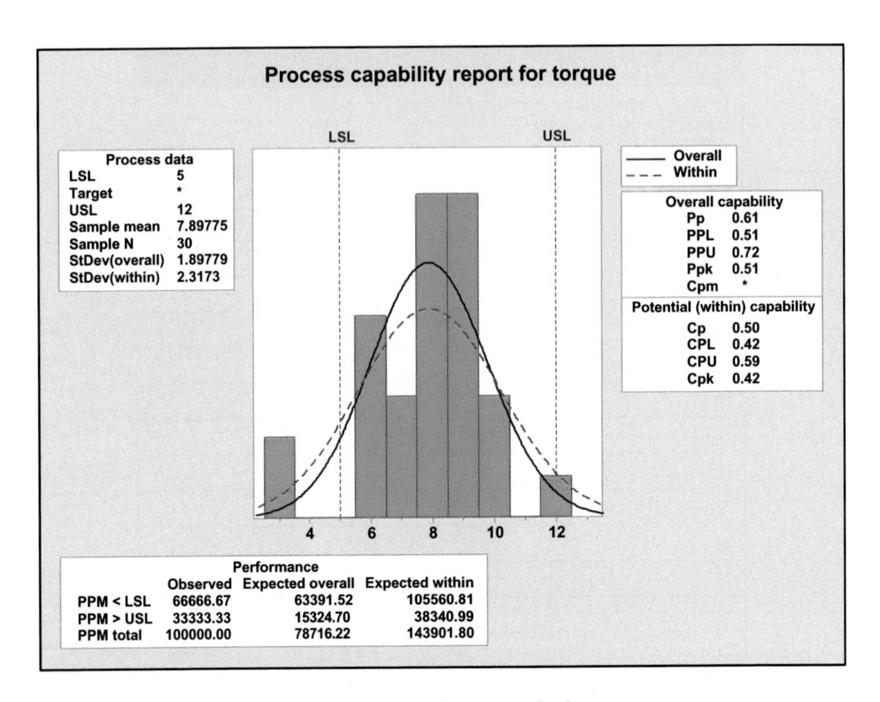

Figure 6.5 Large volume parenteral product process capability example plot.

variability of attributes or parameters early on and set scientifically sound and achievable specifications which will be met by the stable and capable processes.

Ninth Concept—Hypothesis Testing

One of the important types of statistical inference is an assessment of the provided evidence (typically from results of testing or observation readings) in favor of a certain claim about populations. For instance, these analyses determine if two or more populations are equivalent to each other. Therefore a statistical hypothesis which is also called a significance testing is confirmatory data analysis where a hypothesis is examined which based on process observations derives that process is modeled from a set of random inherent variables. Typically, two or more statistical data sets are compared, or a data set obtained by sampling is compared against a synthetic data set from an idealized model or from data attained in different stages of Process Validation. We will examine an example of FDA Form 483 Observation where FDA recommends usage of such analyses. This example, although not citing parenteral product manufacturing and analysis, is also significant because it discusses case of inappropriateness of previously discussed concept eight (capability analysis) use.

In this case four tablet products of various strengths during routine production were manufactured using double-sided tablet press, even though an initial process qualification used a single-sided tablet press. The variability of compression using the double-sided press was not qualified and the firm claimed equivalency of products' populations compressed from both sides by referencing Cpk values for processes using a double-sided tablet press and the single-sided tablet press (this analysis was submitted in firm's respond to FDA Form 483). Upon evaluation of firm's respond FDA had issued a warning letter were amongst other reason for issuance stated (McNally, 2015) that "the Cpk value alone was not an appropriate metric to demonstrate statistical equivalence. Cpk analysis requires a normal underlying distribution and a demonstrated state of statistical process control. The firm did not address these issues in their response. Statistical equivalence between the two presses could have been shown by using either parametric or non-parametric (based on distribution analysis) approaches and comparing means and variances. Neither of these approaches was employed. Firm did not use the proper analysis to support their conclusion that no significant differences existed between the two compression processes."

In other words, FDA had outlined the following issues:

- Data did not support proper statistical conclusions.
- Firm did not understand underlying assumptions required to conduct Process Capability calculations.
- Firm did not conduct proper statistical analysis to demonstrate equivalence between two operations.

The proper analysis in this situation would have been a hypothesis testing such as 2-sample *t*-test. Therefore it is an important tool for specific situations which would allow practitioners to determine inter (within the lot) and intra (between the lots) variability.

Tenth Concept—Design of Experiments

The last concept we would like to bring to the attention of a Process Validation practitioner is a concept of a robust design or a design of experiments. This concept is should have been brought up first because it should be laying a base of any capable and stable process. However, for years, and even now, robust design of experiments is not being used by research and development organizations. In 2004 FDA's Final Report on CGMP's for 21st century regulators inspirationally wrote—"to implement its vision, ICH established two Expert Working Groups (EWGs) on pharmaceutical development. The first (ICH Q8 EWG) seeks to incorporate elements of risk and quality by design throughout the life cycle of the product. The ICH Q8 EWG articulated the 'desired state' for pharmaceutical manufacturing in the 21st century as: Product quality and performance achieved and assured by design of effective and efficient manufacturing processes."[6] The quality by design (QbD) model is being encouraged by global regulators. For instance, in March 2011, the EMA and FDA launched a pilot program to perform a simultaneous joint but separate evaluation of QbD (Quality by Design) elements, as a pilot program. They even issued Q & A document detailing some of reached agreements on a wide range of QbD components (EMA/603905/2013, 2013). Regulators recommend research and development organizations to "design and conduct studies (e.g., mechanistic and/or kinetic evaluations, multivariate design of experiments, simulations, modeling) to identify and confirm the links and relationships of material attributes and process parameters to drug substance CQAs." (ICH, 2012) Once again, we will not go into a detail of regulators recommendations or present technical recommendations on how to perform statistically sound experiments; however, we will list a few necessary rules one needs to follow to implement this concept. First of all one should remember that Design of Experiments (DoE) is the purposeful changes of inputs to a process in order to observe the corresponding changes in the outputs. It is a scientific approach to gain process knowledge; therefore, failure of some DoE batches typically beneficial for process understanding. In other words, DoE is an experimental design that allows to study of both main effects and the interaction between factors on the outputs under study. This is achieved because factors are varied simultaneously rather than varying "one factor at a time." A careful DoE planning that is required to ensure that one gets the most out of a DoE include the following points to consider and rules:

[6] Pharmaceutical CGMPS for the 21ST century—A risk-based approach, final report, Department of Health and Human Services, U.S Food and Drug Administration, September 2004.

1. Define the problem: Developing a good problem statement helps ensure you are studying the right variables.
2. Define the objective: A well-defined objective will ensure that the experiment answers the right questions and yields practical, usable information.
3. Develop an experimental plan: Do research, analyze available data (e.g., Multi-Vari analysis), discuss with experts and interview operators to identify which factors or process conditions affect process variability and performance.
4. Define the factors and factor levels: Partition the key variables into fixed (hold these constant), experimental (the factors you will vary) and noise (factors you must be aware of but cannot vary). Establish the levels for experimental factors and "push" them within safe limits to encourage a response.
5. Create an appropriate experimental design: Designs include screening designs, fractional factorial and full factorial approaches.
6. Determine the number of replications: Replication is the collection of more than one observation for the same set of experimental conditions. Replication allows an estimate of experimental error and the impact of factors on variability to be studied. Replication also serves to decrease bias due to uncontrolled factors.
7. Make sure the process and measurement systems are in control. Use SPC to establish the process is under control, if it is not under control, look for and address assignable causes first. If the process is not under control the results of your experiment may not be able to be reproduced and can yield erroneous conclusions. Make sure the measurement system is capable by performing an Measurement System Analysis (MSA). If the variation of the measurement system is greater than the effects of the variables, the experiment will not yield useful results (30% is a maximum, 20% is desirable)
8. Apply randomization to performance of experiment to eliminate bias from the experiment, variables not specifically controlled as variables should be randomized. This includes samples, operators, shifts, process parameters, etc. Randomization helps to assure valid estimates of experimental error.
9. Analyze the results, select the optimal factors that meet your objectives and perform a confirmation run: A confirmation run involves running the process at the recommended factor levels and verifying the predicted results are achieved.

Conclusion

This concludes our brief recommendation for main concepts to be used for a Risk-Based Life Cycle approach to Process Validation for Parenteral Products manufacturing. We would like to reiterate that the major rule of thumb for use of statistics is to utilize proper tools for different stages of a Process Validations and different cases to be analyzed, such that different tools are used for different data types and different questions, as required. Some of these different "types" are listed below:

- Time ordered data analysis (during the run);
- Batch data analysis (population as a whole);
- Intrabatch variability (within in same batch);
- Interbatch variability (same product different batch);
- Process performance (A measure of actual results achieved by following a process);
- Process capability (The range of expected results that can be achieved by following a process).

In the guidance on Process Validation FDA "recommend(s) that a statistician or person with adequate training in statistical process control techniques develop the data collection plan and statistical methods and procedures used in measuring and evaluating process stability and process capability" as well as other statistics of the process. We are recommending a statistical training for a wide range of disciplines. We believe that it is important to understand variability of the process for all departments involved in a Risk-Based Life Cycle Approach for Process Validation. The depth and the breadth of these statistical training sessions should commensurate a disciplines role in the overall process and should be established by firms, but some level of elementary statistical understanding should be embedded into progressive thinking forward effective organization.

We also would like to briefly mention several statistical software programs that are typically and historically used by many pharmaceutical and biopharmaceutical organizations. The goal of this listing is not to be all inclusive or voice any preference but simply mention statistical software programs an author most frequently encountered in this industry. The firms that developed or distributed these programs are listed as they were at the time of drafting of this chapter (Table 6.3).

Table 6.3 Statistical Software Packages

Software Package	Developer or Distributor
Minitab	Minitab, Inc.
Microsoft Excel	Microsoft Corporation
JMP	SAS Institute
SAS	SAS Institute
Statistica	Tibco
Discoverant	Biovia

Finally we wanted to say a few words about a future use of statistics in a practice of Process Validation. We certainly hope that our process models will result in more predictive, stable and capable processes. We are happy to note that more and more firms across the globe are utilizing statistical methods and analyses. There are still making a lot of mistakes in their implementations that sometimes result in regulatory observations, but we hope that they learn from their mistakes and progress development of effective and efficient processes, as well as manufacturing of consistent products that result in a value for our patients.

References

ASTM E2281, Standard Practice for Process and Measurement Capability Indices.

EMA/603905/2013, 2013. Questions and Answers on Design Space Verification, 24 October 2013.

FDA's Guidance for Industry, Quality Systems Approach to Pharmaceutical Current Good Manufacturing Practice Regulations. Available at: <http://www.fda.gov/Drugs/GuidanceComplianceRegulatoryInformation/Guidances/default.htm>.

FDA Guidance for Industry, 2004a. Sterile Drug Products Produced by Aseptic Processing—Current Good Manufacturing Practice, U.S. Department of Health and Human Services, Food and Drug Administration, Center for Drug Evaluation and Research (CDER), Center for Biologics Evaluation and Research (CBER), Office of Regulatory Affairs (ORA), September 2004, Pharmaceutical CGMPs.

FDA Guidance for Industry, 2011. Process Validation: General Principles and Practices, U.S. Department of Health and Human Services, Food and Drug Administration, Center for Drug Evaluation and Research (CDER), Center for Biologics Evaluation and Research (CBER) and Center for Veterinary Medicine (CVM), January 2011, Current Good Manufacturing Practices (CGMP) Revision.

GHTF/SG3/N99-10:2004, 2004. FINAL DOCUMENT, Quality Management Systems—Process Validation Guidance, second ed. SG3, The Global Harmonization Task Force, January 2004.

ICH, 2009a. Harmonised Tripartite Guideline, Pharmaceutical Development, Q8(R2), Current Step 4 version, dated August 2009a, ICH.

ICH, 2009b. Harmonised Tripartite Guideline, Pharmaceutical Quality System, Q10, dated April 2009, ICH.

ICH, 2012. Harmonised Tripartite Guideline and Manufacture of Drug Substances (Chemical and Biochemical Entities and Biotechnological/Biological Entities) Q11, Current Step 4, version dated 1 May.

Kelly, D., 2017. Celltrion Receives FDA Form 483 Noting 12 Inspection Observations, The Center for Biosimilars, 7 September 2017. <http://www.centerforbiosimilars.com/news/celltrion-receives-fda-form-483-noting-12-inspection-observations>.

Kotz, S., Johnson, N.L., 1993. Process Capability Indices. Chapman and Hall, London.

McCoy, P.F., 1991. Using performance indexes to monitor production process. Qual. Prog. 24 (2), 49–55.

McNally, G.E., 2015. Food and Drug Administration Center for Drug Evaluation and Research Office of Pharmaceutical Quality Office of Process and Facilities, Lifecycle Approach to Process Validation Presentation, FDA_PDA Joint Regulatory Conference, Washington, DC, 28–30 September 2015.

Moore, D.S., McCabe, G.P., 2005. Introduction to the Practice of Statistics, fourth ed. Purdue University, W. H. Freeman and Company, New York.

Taguchi, G., 1995. Quality engineering (Taguchi methods) for the development of electronic circuit technology. IEEE Trans. Reliab. IEEE Reliab. Soc. 44 (2), 225229. Available from: https://doi.org/10.1109/24.387375. ISSN 0018-9529.

World Health Organization, 2005. WHO Technical Report Series, No. 929, WHO guidelines for sampling of pharmaceutical products and related materials, Annex 4.

Further Reading

FDA, 2004b. Report Pharmaceutical CGMPS for the 21st Century—A Risk-Based Approach, Final Report, Department of Health and Human Services, U.S Food and Drug Administration, September 2004b.

Minitab Statistical Software, ver. 17, Help “Normal Distribution”.

National Institute for Standards and Technology Tutorials: <https://www.nist.gov/sites/default/files/documents/2017/05/09/robdesgn-1.pdf>.

CHAPTER 7

Process Validation Stage 1: Parenteral Process Design

Igor Gorsky
Senior Consultant, ConcordiaValsource LLC., Downingtown, PA, United States

Introduction Into Pharmaceutical Development

In the beginning of the new millennium FDA regulators embarked on the journey which would include providing pharmaceutical and biopharmaceutical industries with a frame of science based and risk-based approach to validation of processes, namely processes that involve manufacturing of the product. In a mind of regulators this approach would nourish a culture of robust design of the processes and continued learning about them which would sustain a principle of knowledge management. In November 2003, ICH (International Council for Harmonisation of Technical Requirements for Pharmaceuticals for Human Use) started working on a harmonized plan which included a development of a "pharmaceutical quality system based on an integrated approach to risk management and science. To implement its vision, ICH established two Expert Working Groups (EWGs) on pharmaceutical development. The first (ICH Q8 EWG) seeks to incorporate elements of risk and *quality by design* throughout the life cycle of the product. The ICH Q8 EWG articulated the 'desired state' for pharmaceutical manufacturing in the 21st century as: Product quality and performance achieved and assured by design of effective and efficient manufacturing processes" (FDA, 2004).

The main concentration of ICH Q8 working group was on process design which would provide guidance to formulators to work closely with validation, operations, and quality assurance groups on obtaining early information on process capability and variability from process development efforts, as well as mounting and monitoring data to attain knowledge and more importantly understanding on process variability and relationships between the process parameters and product quality

Principles of Parenteral Solution Validation. DOI: https://doi.org/10.1016/B978-0-12-809412-9.00006-X

resulting in robust process control strategies that would allow manufacturers to be proactive and resolve issues armed with this knowledge. Early in the process development formulators and research and development staff would conduct "experiments and studies to identify and establish process parameter relationships and sources of variability" based on risk assessment that may be "used to minimize and prioritize efforts" (FDA, 2004), as it could become cumbersome and ineffective. Therefore the use of prior knowledge should be widely utilized and in fact provide a platform on which future processes should be build. "Invention of the wheel" should be prevented as much as possible. However, it is recognized that new technologies may lead to use of new process solutions for new products. It is especially recognized that rise of the importance of cell therapy products have shown to bring about evolution of the new complex solutions and methods that include wide use of Design of Experiments (DoE), as purification processes require multiple levels of chromatography and filtration. In addition, concerns about an aseptic processing of commercial production bring another layer of complexity in design and implementation of these processes. The final development of process control strategy prior to Stage 2 is becoming a crucial moment in life cycle of a product and therefore consistent and reproducible process is in a core of the Stage 1 development of design.

There are, of course, a number of guidances available from ICH, FDA, EMA, WHO, and industry professional organizations which are found in the bibliography references. The most important of these guidances is ICH Q8 Pharmaceutical Development Guidance, which serves as a robust pillar for the industry and provides a high level direction on how to perform pharmaceutical development using knowledge and risk management and eventually transfer this body of knowledge to Stage 2 (Process Performance Qualification, PPQ) practitioners as well as supply information for a section 3.2.P.2 (Pharmaceutical Development) of a regulatory submission in the ICH M4 Common Technical Document (CTD) format. Although we are not going to fully recite entire ICH Q8 guideline, we will provide major elements in this chapter. In addition, we will add a case study which outlines an example of the DoE for fermentation process for parenteral vaccine along with setup, experiment review and interpretation of results. This case study should help reader to understand value of the DoEs which leads toward quality by design (QbD) aspirations of industry and regulators.

Master Planning, Organization, and Schedule Planning

The Master Planning is one of the most activities that should be done prior to starting any product/process development. Typically, effective governance of master planning systems is achieved through establishing policies or standards documents that describe in general terms, the requirements and elements of systems and how it will be implemented. These documents may not have step by step instructions for how to perform operational tasks but should further breakdown the parts of the regulatory guidance's into specific subjects or topics in relation to firm's product/process development and validation needs. Fig. 7.1 shows major elements of master planning that include validation activities and enablers/deliverables. As shown in Fig. 7.1, on the left side there is a list of enablers/deliverable documents/reports and on the right side are standard activities for process validation stages of a parenteral drug product.

Red (low), yellow (medium or moderate), and green (high) colors represent a level of knowledge we typically possess at different stages of Process Validation. During the life cycle of the product we will progress through these stages. Our process knowledge shall be built over time and will help us understanding process variability, as well as help guiding us in decision-making and manufacturing management. In addition, this knowledge will help us developing processes for other parenteral products.

Points to consider in Master planning:

1. First and foremost, we need to remember that a Master Plan is a living document. It should be designed with agility to make changes, as they may occur. Parenteral products process developers should be prepared for multiple changes that may happen due to initially unknown variability of attributes and parameters. Many parenteral products are cell therapies or vaccines which require growth and promotion of cells that may cause a high level of variability in the beginning of the development process. In addition, one may require changes to equipment, facilities or utilities upon gaining understanding of variances of the process and its outcomes. Therefore living document status of a Master Plan would be helpful to scientists to be able to document and follow original design through its changes and to measure a level of success from the beginning of development to the level at which deeper knowledge of process is obtained.

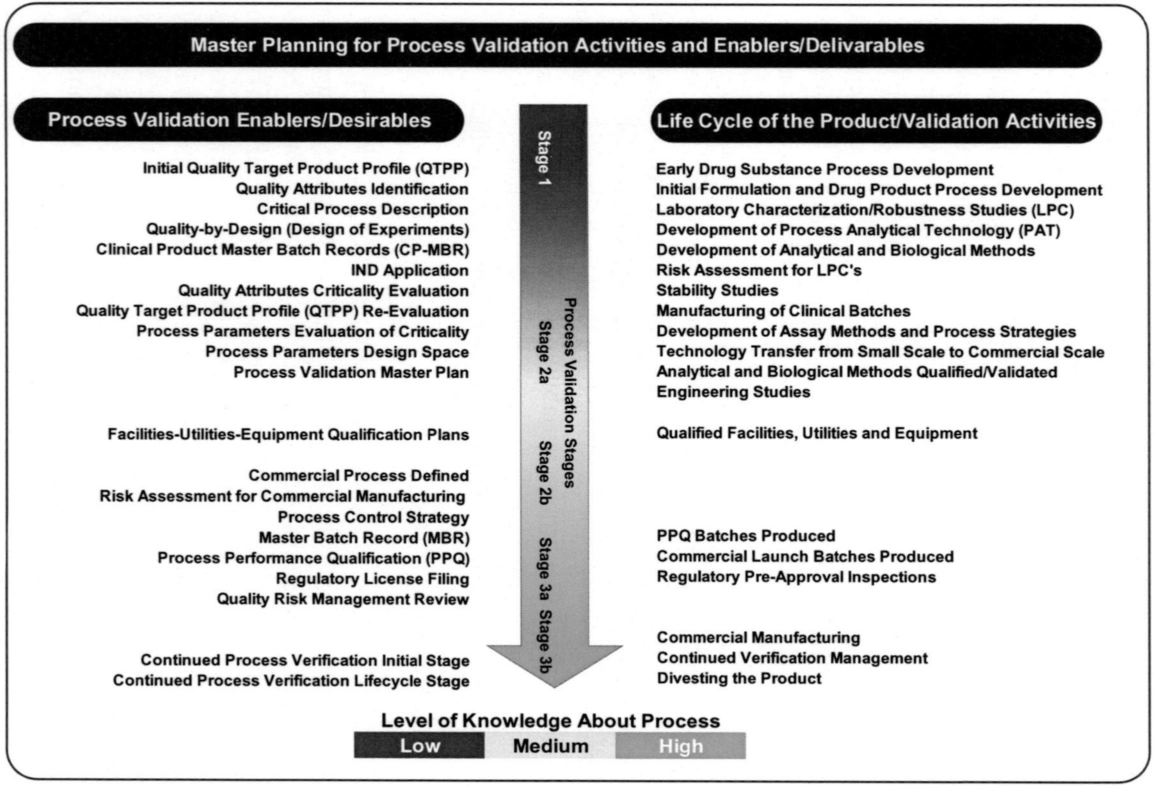

Master Planning for Process Validation Activities and Enablers/Delivarables

Process Validation Enablers/Desirables

Initial Quality Target Product Profile (QTPP)
Quality Attributes Identification
Critical Process Description
Quality-by-Design (Design of Experiments)
Clinical Product Master Batch Records (CP-MBR)
IND Application
Quality Attributes Criticality Evaluation
Quality Target Product Profile (QTPP) Re-Evaluation
Process Parameters Evaluation of Criticality
Process Parameters Design Space
Process Validation Master Plan

Facilities-Utilities-Equipment Qualification Plans

Commercial Process Defined
Risk Assessment for Commercial Manufacturing
Process Control Strategy
Master Batch Record (MBR)
Process Performance Qualification (PPQ)
Regulatory License Filing
Quality Risk Management Review

Continued Process Verification Initial Stage
Continued Process Verification Lifecycle Stage

Process Validation Stages

Stage 1
Stage 2a
Stage 2b
Stage 3a
Stage 3b

Life Cycle of the Product/Validation Activities

Early Drug Substance Process Development
Initial Formulation and Drug Product Process Development
Laboratory Characterization/Robustness Studies (LPC)
Development of Process Analytical Technology (PAT)
Development of Analytical and Biological Methods
Risk Assessment for LPC's
Stability Studies
Manufacturing of Clinical Batches
Development of Assay Methods and Process Strategies
Technology Transfer from Small Scale to Commercial Scale
Analytical and Biological Methods Qualified/Validated
Engineering Studies

Qualified Facilities, Utilities and Equipment

PPQ Batches Produced
Commercial Launch Batches Produced
Regulatory Pre-Approval Inspections

Commercial Manufacturing
Continued Verification Management
Divesting the Product

Level of Knowledge About Process

Low | Medium | High

Figure 7.1 Master planning for process validation activities and enablers/deliverables.

2. A Master Plan is a blue print for the project. Just like an architectural or engineering, blue print is essential for construction of a building, so is a Master Plan essential for successful process design.
3. A Master plan must provide a system and process definitions to assure that there is a good understanding of unitary whole and its components. An entire process may be considered as a system, while subcomponents are its processes. A system and each of its subelements (processes) should be succinctly described in a master plan including facilities, utilities, and equipment required in its production.
4. A Master Plan must also have a format which is easy to present, be trained on and follow. It should easily adhere to organize and plan approach which may include the following elements:
   a. clear expectations;
   b. defined scope;
   c. process plan;
   d. project management elements;
   e. detailed schedule;
   f. risk management elements;
   g. stage-gate reviews against plan;
   h. efficient and effective resolution of deviations; and
   i. Multidiscipline expertise.
5. Management must be involved and buy into a Master Plan early in the development cycle. Install into development stage-gate reviews and updates with upper management. Specifically, resources, schedules, and budgets must commensurate risk assessments performed and knowledge available at every stage of the pharmaceutical development. Rationales for Laboratory Characterization studies, Engineering batches, and Clinical batches should be also based on risk assessments and knowledge attained at each stage of development.
6. Master planning should consider PPQ Plan, as well as Continued Process Verification plans early on and update them based on the knowledge gained. Smartly evaluated and used knowledge may save time and money during implementation of subsequent stages of Process Validation.

Risk/Impact Assessment

As the approach described in this book is science- and risk-based this task is essential prior to implementation of laboratory/bench characterization studies. It must be done methodically following determination of Quality Target Product Profile (QTPP). An importance of defining QTPP early on is essential as it "forms the basis of design for the development of the product" (ICH, 2009). While developing the QTPP one may consider the following elements:

- intended use of the product in clinical or therapeutic environment;
- products route of administration which in a case of a parenteral drug is an injection;
- products dosage form which for parenteral medicine is a vial;
- product delivery systems that are methods by which drug has been administered to achieve intended therapeutic effectiveness, that is, pulmonology, nasal delivery, blood stream, brain, eye ball, etc.;
- drug product dosage strength(s);
- a container closure system;
- therapeutic moiety release or delivery;
- attributes affecting pharmacokinetic characteristics (e.g., therapeutic performance);
- drug product quality attributes criteria, such as:
  - sterility,
  - purity,
  - efficacy,
  - safety,
  - stability, and
  - drug release.

Often companies do not spend enough time to define QTPP at an early stage of development which may cause ambiguity and unnecessary studies that may not provide relevant process development information. Upon defining of an early QTPP a multidisciplinary team may continue to perform a Risk Impact Assessment. We will go over a short description of a Risk Impact Assessment. Fig. 7.2 illustrates a Risk Management process as it is presented in ICH Q9 Guidance for Quality Risk Management. It is important to understand a definition of "Risk." It is generally defined as a hazard to the patient, process, compliance, or business. Risk is typically measured by multiplying a severity of a hazard by the probability of knowledge of occurrence. In addition, a level of detectability has been historically used as a mitigation factor in risk evaluation process. Sometimes the level of detectability for the pharmaceutical development is hard to determine in the

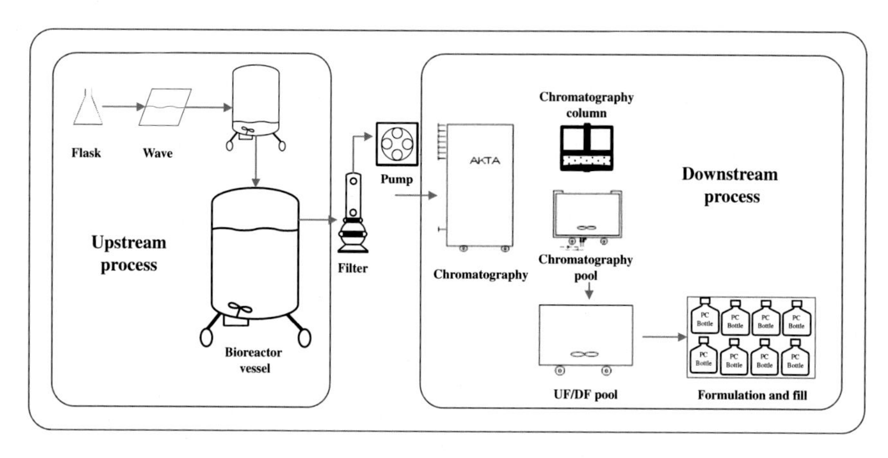

Figure 7.2 Vaccine/cell therapy product simplified process.

early stages of development; however, one may need to resort to scientific and technical literature or rely on the previous knowledge. The objective of a Risk Impact/Assessment is to develop a manufacturing process that would consistently meet its predetermined criteria. As shown in Fig. 3.1 Risk Management consist of the following steps:

Risk Assessment that should assist in understanding of attributes and discovery of the impact of parameters on the attributes:

- Risk Identification—determination of harms and/or hazards to patient, product, process, compliance, or business.
- Risk Analysis—determination of causes of the harms or hazards.
- Risk Evaluation—evaluation, ranking or scoring of the risks (harms or hazards), where detection level may play a significant role.
- Risk Control—determination of actions that must be taken to control risks.
- Risk Reduction—identification of improvement opportunities to control risks.
- Risk Acceptance—deciding to live with those risks that may not be reduces or controlled but will monitored.
- Risk Communication—interaction with the discipline, between the discipline and upper management regarding risk management decisions.
- Risk Review—follow-up and feedback for risk management.

It should be noted that Risk Management exercises are performed by multidiscipline groups where at minimum representatives of

Research and Development (Product sand Process Development), Operations, Quality Control, Engineering, Technical Services, Regulatory Assurance, Statistics, and Quality Assurance should be present. A high level of Aseptic Processing understanding, and expertise should be present in a group to assure that decisions made for parenteral product do not jeopardize sterility of the product. Finally these events are not on–offs and should be continually exercised.

Process/System Design

Since parenteral products could be solutions, suspensions, emulsions for injection or infusion, powders for injection or infusion, gels for injection, and even implants a design of their processes may vary dramatically. The similarities of the process/system design would be in the area of aseptic processing and ultimate sterility of the product since all of these products are intended for direct administration into the blood circulation of the human or animal body. Therefore we shall only list major points to consider during the risk-based review of the design.

1. Product Understanding:
   One must answer the following questions to explore product understanding:
   What defines the product as useful? This perhaps could be answered when drafting a QTPP. In addition, one should always keep in mind product's purpose to continually assess the risk to the patient.
   What are the critical quality attributes (CQAs) of the product? Once again, a QTPP could be useful for answering this question. We will also describe methods for CQAs determination further in this chapter.
2. Process Understanding:
   Next step would be understanding the process by which product recommended to be produced by. Elementary question for this task would be:
   What are the process steps which affect the critical quality attributes of the product? For instance, typical steps for vaccines, cell/genetic therapy products would include upstream and downstream processes operations as shown in simplified Fig. 7.2.
   An Upstream Processing would typically include steps from initial thaw of biomaterials through sometimes multiple steps in which biomolecules/cells are grown using vials, waves, and bioreactors.

Special media and bacterial or mammalian cells are used to promote and sustain the growth until appropriate yield/cell density is reached. Upon completion of the desired yield production the product is harvested and moved onto downstream part of the overall process. A Downstream Processing typically consist of further purification of products mainly using various chromatography methods and other techniques removing undesired residues, such as tissue or fermentation broth, for instance and filtering them including microbial, endotoxin, and virus filtrations. Disposal of undesired waste is also part of the downstream process.

What are the conditions, we wish to avoid, the unwanted conditions or attributes? Based on the review of the quality attributes and process steps relationship to them, strategies to avoid the unwanted conditions or attributes outcomes could be examined and preliminary understanding could be developed.

3. Risk Analysis

   A risk analysis may include the questions like:
   - What is the impact of the loss of a CQA?
   - What is the probability or likelihood of the occurrence of the failure?
   - What is the chance that you will detect the failure? And if you do, can you correct it?

   We will discuss this in further sections which discuss more details of risk assessment.

4. Risk Acceptance Decision

   A risk acceptance decision is an important step in Process/System Design. When based on the risk assessment, analysis and evaluation reduction may not be feasible, one may decide to accept the risk. If risk acceptance is an option how can we accept the risk? Is it possible trying to reduce it? What should be taken into considerations when making these decisions? This may depend on:
   - product usage;
   - confidence in process "robustness"; and
   - regulatory and quality history.

5. Risk Reduction

   How do we reduce risk? Risk may be reduced by decreasing probability of occurrence, increasing chance of hazard detection or ...by changing (improving) the process. Many times risk reduction requires thinking "out of the box," introduction of new technologies or some other bold steps to prevent hazards from occurring.

6. Follow-up
   Once decisions with regards to process/system design are taken we must communicate this information to appropriate teams, as well as upper management. At this point decision regarding risks are part of the change management and are subject to periodic or stage-gate reviews. Special attention should be given to a residual risk or a possibility of residual risk.

Perform Risk Assessment (Identification of Critical Quality Attributes and Critical Process Parameters)

We already talked about principals of a risk assessment. Now we will discuss details of an actual process. A risk assessment is which quality attributes and process parameters are discussed should be a "classroom" multidisciplined exercise. We will discuss major principals of this exercise and will give an example of an outcome worksheet that should be documented in the Risk Assessment report.

First, attributes including In-Process Specifications (IPCs) that are typically used for characterization should be listed. A worksheet is an excellent tool for such a list. Typically attributes are listed as columns on this worksheet and unit operations that are subsequently broken into subprocesses and process parameters are listed as rows.

Some of the prerequisites based on our previous knowledge should include the following considerations:

1. The process and explanation of defining criticality should be documented in the report.
2. The process should include explanation on expectations with regards to residual risks and what controls are in place to prevent or reduce them.
3. The relative criticality of the CQAs may not be provided, as criticality is a continuum (PDA, 2012; FDA, 2011) which requires risk assessment to be performed such as Failure Mode and Effect where critical process parameter (CPP) versus CQAs relationships are gauged using statistical justifications.
4. CQAs criticality should be determined based on the potential patient impact, for instance.
5. CQAs—trace matrix should be correlated with QTPPs.
6. Critical Material Attributes should be considered.

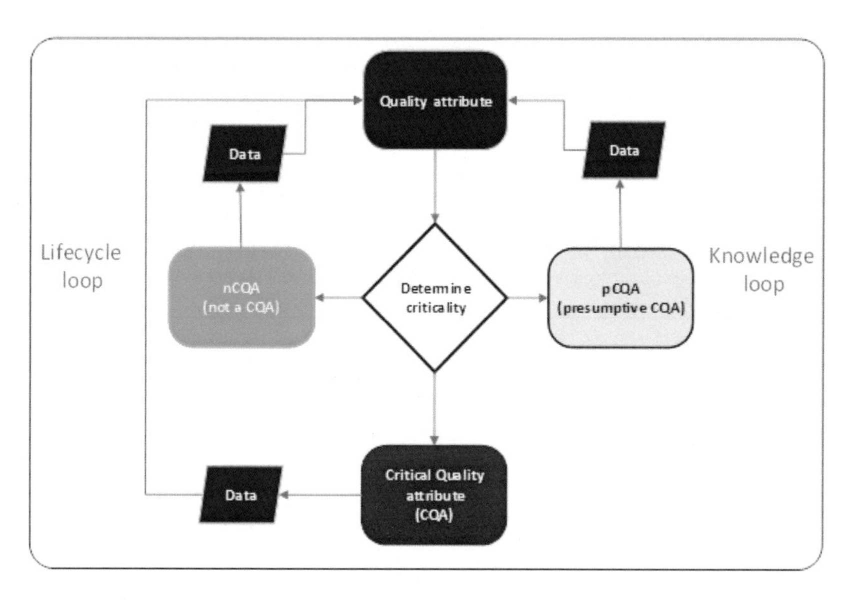

Figure 7.3 Quality attributes criticality assessment process.

7. CPPs may need to be sieved through as typically there are too many categorizations. One must take into consideration Pareto principle where 80% of the process is controlled by 20% of the parameters. It is most likely a case, as we too many parameters variability would be hard to control. For instance, mixing time is a critical parameter which impacts CQAs, therefore it may not have a key Performance Parameter (kPP) designation.
8. The relative criticality of the CPPs typically may not be provided because it is also a continuum.

An example of an assessment of a Quality Attribute criticality is shown in Fig. 7.3 decision tree process. One starts with what is identified as a quality attribute—that is typically "a physical, chemical, biological, or microbiological property or characteristic that should be within an appropriate limit, range, or distribution to ensure the desired product quality (ICH, 2009)." One's decision regarding criticality of the attribute should be based on knowledge, as well as understanding of process shift over life cycle of the product. As stated earlier, the criticality is a continuum and it is possible that over time of the product life, as the product knowledge builds criticality of attributes may change.

Table 7.1 Hazard Severity—Keywords and Descriptions

Severity	Patient/Product/Process Criticality
Critical	Serious Adverse Experience. Will cause permanent impairment or damage of a body structure or function. Could lead to patient death
Major	Serious injury. Could cause permanent impairment or damage to a body structure or function but is not fatal or life threatening. Likely to impact product efficacy
Moderate	Nonserious injury. May cause significant but recoverable injury to the patient or user. May cause significant temporary unintended impairment of a body function. May impact product efficacy
Minor	May cause transient, self-limiting, unintended, impact to a body function. May cause dissatisfaction to the patient and customer complaint
Negligible	No performance impact to patient. May have cosmetic defect which is unlikely to cause dissatisfaction to the patient

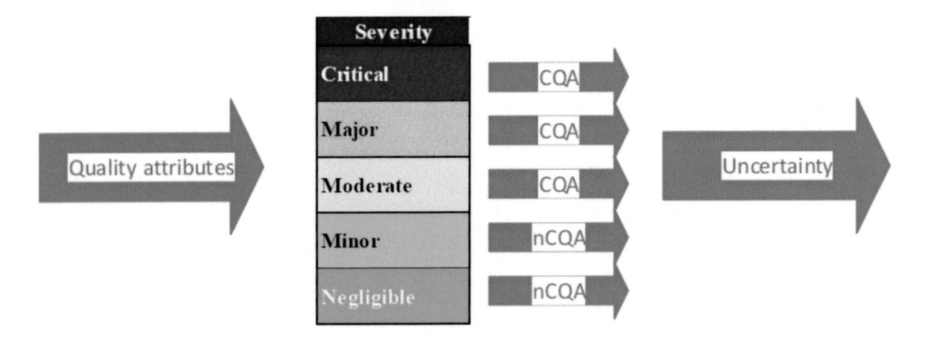

Figure 7.4 Evaluation of quality attributes severity.

Similarly, to the approach shown in Chapter 3, Quality Risk Management, the severity of quality attributes may be assessed first using the following types of key words and descriptions shown in Table 7.1.

Upon establishing keywords and description procedure Quality Attributes criticality assessment following the process shown in Fig. 7.4.

As shown in Fig. 7.4 those quality attributes that scored as Minor or Negligible on Severity scale may not be CQAs. An example of comparison of these descriptions and criticality determination with higher Severity rankings is shown in Table 7.2.

Upon determination of Severity an assessment with regard to Uncertainty follows, as we remember that Risk is measured by

Table 7.2 Quality Attributes Criticality Comparison and Analysis

Attribute	Severity Level of Potential Harm	Uncertainty	Justification	Review Process
Sterility	Critical (5)	Low	Parenterally delivered drug product containing microorganisms even in low dose may result in systemic infection	The severity level of potential harm to the patient of the quality attribute is rated catastrophic
				Based on regulatory expectations this quality attribute needs to be tested and controlled
Protein concentration	Major (4)	Low	The severity evaluation considered scientific data that indicates a linear relationship exists between concentration and potency as measured by the bioactivity assay. Dose dependence is also demonstrated in animal models and by human clinical data	Severity is scored as significant due to the potential severe impact to efficacy. Based on regulatory expectations this quality attribute needs to be tested or controlled

Table 7.3 Uncertainty Levels Description

Level	Description
High	Data are not available. The severity level is based on limited scientific literature on similar processes/products in combination with scientific judgment
Medium	The severity level is based on in-house scientifically derived limited data in combination with scientific literature on related processes/products
Low	The severity level is based on extensive in-house scientifically derived data consistent with the literature on related products or the severity level is based on strong scientific rationale supported by extensive literature

Severity multiplied by Occurrence (in this case Uncertainty, as we may not yet know a level of occurrence). Table 7.3 lists the Uncertainty descriptions that may be used during Stage 1 process.

Once the descriptions are identified one can start scoring Quality Attributes using both Severity and Uncertainty scales which are illustrated in Table 7.4.

An example of Quality Attributes scoring is shown in Table 7.5.

Next in the Risk Assessment exercise impacts of Quality Attributes on Unit Operations are examined. Fig. 7.5 shows an

Table 7.4 Determination of Quality Attributes Using Severity and Uncertainty Rankings

		Uncertainty		
		Low	**Medium**	**High**
		Large amount of in-house knowledge, large body of knowledge in literature	Some in-house knowledge and scientific literature	No/little in-house knowledge, very limited information in scientific literature
Severity	High (catastrophic patient impact)	Critical	Critical	Critical
	Medium (moderate patient impact)	Potential	Potential	Potential
	Low (marginal patient impact)	Noncritical	Noncritical	Noncritical

Table 7.5 Example of Quality Attributes Scoring

		Uncertainty		
		Low	**Medium**	**High**
Severity	Critical	CQA	CQA	CQA
	Major	CQA	CQA	CQA
	Moderate	pCQA	pCQA	CQA
	Minor	nCQA	pCQA	pCQA
	Negligible	nCQA	nCQA	pCQA

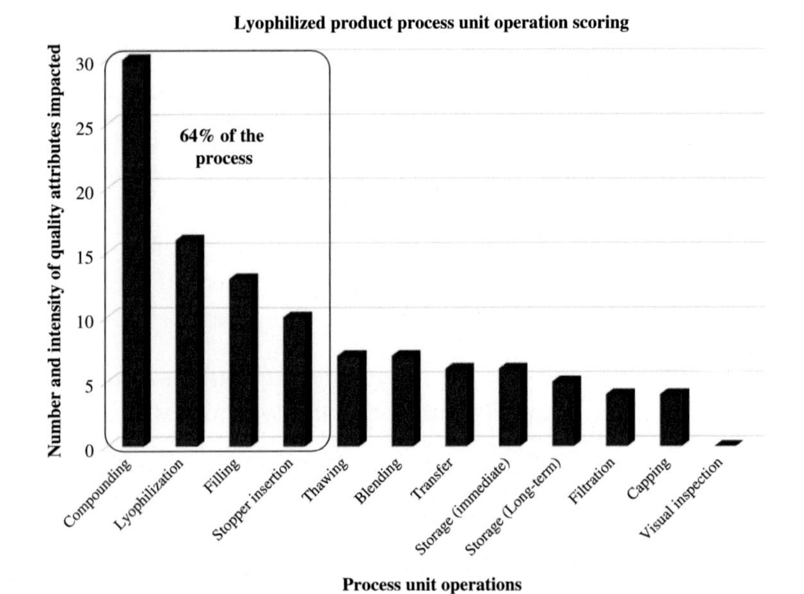

Figure 7.5 Example product unit operations scoring of quality attributes.

example bar chart of Quality Attributes impacts on Unit Operations for a typical lyophilized product. As shown in this chart a Risk Assessment had determined that four-unit operations (Compounding, Lyophilization, Filling, and Stopper Insertions) have approximately 64% impact of all quality attributes. Since a Risk Assessment determined in this example that these unit operations have the most criticality, a decision could be made to further investigate these unit operations using DoEs.

Design of Experiments

The statistical DoE as a powerful tool is encouraged to be used during Stage 1 Process Design. The goals of DoE shall include:

- Determine which process input parameters have a significant effect on the output quality attributes.
- Help determine the "design space" levels of the input parameters that will produce acceptable output quality attribute results.
- Optimize the output quality attributes, such as yield and acceptable levels of impurities.
- Determine levels of the input parameters that will result in a robust process that reduces process sensitivity to variability in process parameters.
- DoE differ from the classical approach to experimentation, where only one parameter is varied while all the others are held constant. The "one-factor-at-a-time" type of experimentation may not efficiently determine process parameter interactions, where the effect of one parameter on a quality attribute differs depending on the level of the other parameters. The basic steps for the DoE approach are summarized below.

Determine the input parameters and output quality attributes to study. This is best done as part of a team approach to identify potential critical process parameters and quality attributes, as discussed in previous section on Risk Assessment. In many cases, the process may be well understood and the parameters and attributes for experimentation readily determined. If there are a large number of input parameters to study, an initial screening design such as a fractional factorial design may be used. The purpose of a screening experiment is to identify the critical parameters that have the most important statistical effect on the quality attributes. Since screening designs do not always clearly identify interactions, the reduced number of parameters identified by the screening experiment will be included in further experiments. If the change is to an existing process, it is often valuable to construct a Multivari chart or statistical process control (SPC) chart from current process data. A Multivari chart can be used to identify if the biggest sources of variation are within-batch variation, between-batch variation, or positional variation (such as between fill heads on a multihead filler). Variance components can also be calculated from the data to determine the largest component of variance. Process parameters are then identified that could be causing the largest sources of variation; these parameters are included in subsequent experiments. Conduct the experiment(s) to determine which parameters have a significant main or interaction effects on the quality attributes. This will usually be a full factorial design for the selected parameters. A factorial design is one in which a low (−) and high (+) level is selected for each factor (parameter). If possible, control runs at the nominal midpoint levels (0) between the low (−) and high (+) levels of the factors should be included in the experimental design. Using control runs at the beginning and the end of the factorial experiment, and ideally also during the factorial experiment, will allow detection of any process drift during the experiments. If possible, the parameter effects on both the mean and variation of the quality attributes should be determined. Some parameters may affect the mean only, variation only, or both. This information can be used to minimize the variation while optimizing the mean, which results in a robust process. Standard DoE approaches may be used for this, but an approach called the Taguchi method is also sometimes used. Optimize with response surface experiments and determine design space. Occasionally the science behind a process will be understood well enough to start with response surface experiments without performing initial screening. Or sometimes enough information will be learned from factorial studies that no additional experiments will be required. However, it is often necessary to conduct more extensive experiments at more than three levels for the parameters identified as the most important from earlier factorial experiments.

To reiterate, the DoEs is the purposeful changes of inputs to a process in order to observe the corresponding changes in the outputs. It is a scientific approach to gain process knowledge and help building

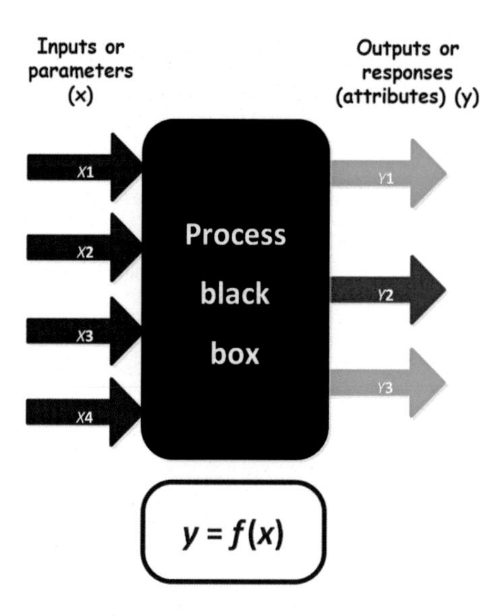

Figure 7.6 Design of experiments concepts.

Quality into the Process (QbD) by studying impacts of process parameters on Quality Attributes. Experimental designs allow the study of both main effects and the interaction between factors on the outputs under study. This is achieved because factors are varied simultaneously rather than varying "one-factor-at-a-time." It is an efficient and effective way to conduct relevant, prospective, and value-added studies. Fig. 7.6 illustrate the concepts of DoEs. The input parameters in our case are CPPs which a varied typically using factorial designs. The responses (CQAs) are measured to assess main effects as well as effect of interactions. At the end of the study the best fitting equation function of CQAs in terms of CPPs is calculated such as shown in Eq. (7.1).

$$Y\,(CQA's) = f(x(CPP's))$$

Equation 7.1 CQA as a function of CPP.

Table 7.6 provides points to consider and prerequisites for DoEs studies. It is important to note that randomization of study runs is a requirement and should never be compromised. It is highly recommended to use statistical software for DoE studies as they provide great tools for analyses of results.

<table>
<tr><th colspan="4">Table 7.6 Points to Consider and Requirements for Design of Experiments and Example of Mixing Study Points to Consider and Design</th></tr>
<tr><th colspan="3">Careful Planning Is Required to Ensure You Get the Most Out of a DoE</th><th>Mixing Study Example</th></tr>
<tr><td>1</td><td>Define the problem</td><td>Developing a good problem statement helps ensure you are studying the right variables</td><td>Current mixing conditions may not be fully defined, and additional experiments are needed to understand impact of mixing on the Quality Attributes and ability of the process to produce uniformed/homogenized product</td></tr>
<tr><td>2</td><td>Define the objective</td><td>A well-defined objective will ensure that the experiment answers the right questions and yields practical, usable information.</td><td>The objective of the study will be to determine the optimum mixing time/speed range at which product mains its homogeneity, as well as to determine mixing impact on protein stability</td></tr>
<tr><td>3</td><td>Develop an experimental plan</td><td>Do research, analyze available data (e.g., Multivari analysis), talk with experts and interview operators to identify which factors or process conditions affect process variability and performance.</td><td>Experimental plan will include studying mixing time and mixing speed to evaluate impact on quality attributes (based on Risk Assessment):
• High Molecular Weight
• Particulate Matter
• Bioburden
• Endotoxin
• Sterility
• Specific Activity
• Content Uniformity
Stability should be assessed</td></tr>
<tr><td>4</td><td>Define the factors and factor levels</td><td>Partition the key variables into fixed (hold these constant), experimental (the factors you will vary) and noise (factors you must be aware of but cannot vary). Establish the levels for experimental factors and "push" them within safe limits to encourage a response</td><td rowspan="3">2 Level, 2 full factorial design is recommended with 1 center block and 2 replicates
<table>
<tr><th>Std. Order</th><th>Run Order</th><th>Center Pt</th><th>Blocks</th><th>A</th><th>B</th></tr>
<tr><td>9</td><td>1</td><td>0</td><td>1</td><td>0</td><td>0</td></tr>
<tr><td>7</td><td>2</td><td>1</td><td>1</td><td>−1</td><td>1</td></tr>
<tr><td>2</td><td>3</td><td>1</td><td>1</td><td>1</td><td>−1</td></tr>
<tr><td>1</td><td>4</td><td>1</td><td>1</td><td>−1</td><td>−1</td></tr>
<tr><td>8</td><td>5</td><td>1</td><td>1</td><td>1</td><td>1</td></tr>
<tr><td>5</td><td>6</td><td>1</td><td>1</td><td>−1</td><td>−1</td></tr>
<tr><td>6</td><td>7</td><td>1</td><td>1</td><td>1</td><td>−1</td></tr>
<tr><td>3</td><td>8</td><td>1</td><td>1</td><td>−1</td><td>1</td></tr>
<tr><td>4</td><td>9</td><td>1</td><td>1</td><td>1</td><td>1</td></tr>
</table></td></tr>
<tr><td>5</td><td>Create an appropriate experimental design</td><td>Designs include screening designs, fractional factorial, and full factorial approaches</td></tr>
<tr><td>6</td><td>Determine the number of replications</td><td>Replication is the collection of more than one observation for the same set of experimental conditions. Replication allows an estimate of experimental error and the impact of factors on variability to be studied. Replication also serves to decrease bias due to uncontrolled factors</td></tr>
</table>

(*Continued*)

Table 7.6 (Continued)

Careful Planning Is Required to Ensure You Get the Most Out of a DoE			Mixing Study Example
7	Make sure the process and measurement systems are in control	Use SPC to establish the process is under control, if it is not under control, look for and address assignable causes first. If the process is not under control the results of your experiment may not be able to be reproduced and can yield erroneous conclusions	Establish methods of data analysis including statistical process control, primary (main effects) and interaction effects, cube plots, etc. Perform multivariate analysis if required
		Make sure the measurement system is capable by performing an MSA. If the variation of the measurement system is greater than the effects of the variables, the experiment will not yield useful results	Perform MSA to assess variability due to method, instruments or analyst
8	Apply randomization to the performance of the experiment	In order to eliminate bias from the experiment, variables not specifically controlled as variables should be randomized. This includes: samples, operators, shifts, etc. Randomization helps to assure valid estimates of experimental error	Refer to table in Items #4–6.
9	Analyze the results, select the optimal factors that meet your objectives and perform a confirmation run	A confirmation run involves running the process at the recommended factor levels and verifying the predicted results are achieved	Determine the best fit equation for optimum parameters to meet Quality Attributes specifications

The goal of response surface experiments is to develop an equation that accurately models the relationship between the input parameters and output quality attributes. This equation is then used to determine the design space region of the input parameters where the output quality attributes will meet specifications. The results of the experiments are analyzed with a computer program such as Minitab, JMP, SAS, etc. to determine regression equations to model the process.

Another aspect of optimization is to develop a robust process. The regression equations already developed can be used to locate input parameter settings that are "forgiving"; that is, when the process is run at these settings, variation in the input parameters will not result in unacceptable variation in the quality attributes. The idea is to stay away from boundaries or areas in the parameter design space where variation in the parameter will result in a rapid deteriorating quality. This is accomplished by using the quadratic and interaction effects to minimize variation. The results may also be used to calculate the percent of total variation attributable to each parameter. This is called a variance components analysis. The input parameters contributing the most to the output quality attribute variation can be controlled the most tightly, made robust by running the process at a particular level of the other parameters, or improved by a process design change to reduce the impact of the parameter.

Developing Control Strategies and Determine Process Design

Once the design space region for the input parameters that results in quality attributes meeting specifications has been determined, it is recommended that additional experiments be used to confirm the expected DoE results. This may consist of running a few experiments at various parameter combinations to verify the DoE equation adequately predicts the results. In some cases where there is already good confidence in the DoE results, the Stage 2 PPQ results may be used for this. Modeling, simulate the commercial process/computer-based or virtual simulations and experiments and demonstrations at laboratory and pilot scale. It is required that activities and studies resulting in process understanding be documented. Documentation should reflect the basis for decisions made about the process. The functionality and appropriateness of commercial manufacturing equipment should be taken into consideration in the process design. Predicted contributions of equipment, different component lots, production operators, environmental conditions, and measurement systems variabilities in the production setting to the overall process variability must be taken into consideration. Process control strategy is the most important deliverable of the pharmaceutical development in Stage 1. The effectiveness of the control strategy shall dictate the extent to which a manufacturing process remains in a state of control, and an appropriate control strategy is based on the knowledge and experience gained in Stage 1.

The development of an effective process control strategy is an iterative process, beginning early in development and evolving as process and product knowledge increase. A robust control strategy encompasses all elements of individual unit operations in the process. All product quality attributes and process parameters whether or not they are classified as critical are included in a complete process control strategy. Elements in a process control strategy shall include but not be limited to:

- Raw Material Controls;
- In-Process and Release Specifications;
- In-Process Controls;
- Performance Parameters;
- Process Parameter Set Points and Ranges;
- Process Monitoring (Data Review, Sampling, Testing); and
- Processing and Hold Times.

Process Analytical Technology (PAT) represents one approach for implementation of the Control Strategy. Using PAT, CQAs are monitored in real time (using online or atline analytics) and results are used to adjust CPPs during production to decrease product variability (CQAs) or to achieve consistent CQAs at desired ranges with low variability. PAT utilizes product and process knowledge as well as equipment automation and analytical instrumentation technologies. Successful application of PAT requires a thoroughly characterized process in which the relationship between CPPs and CQAs is explored using mathematical models such as multivariate analysis. Application of this understanding to the Control Strategy also affects the design and qualification of the instrumentation and control systems in the manufacturing process. In order to support implementation of PAT, Stage 1 deliverables must describe the CQA monitoring scheme and the algorithm for adjusting CPPs based on the process response. Qualification of the equipment, measurement system, and process (Stage 2) must demonstrate the capability to adjust CPPs according to the established algorithm and confirm that these adjustments result in acceptable and predictable outputs.

Scale-Up and Technology Transfer

The basic principles employed in the preparation of parenteral products do not vary from those widely used in other sterile and nonsterile liquid preparations. However, it is imperative that all calculations be accurate and precise. Typically, the issue of parenteral solution scale-up essentially becomes a liquid scale-up task, which requires a high degree of accuracy. There are many practical yet scientifically sound means of performing this scale-up analysis of liquid parenteral systems could be found in the literature. This author developed a method based on the scale of agitation scheme (Parenteral Drug, 2011). For single-phase liquid systems, the primary scale-up criterion is equal liquid motion when comparing pilot-size batches to a larger, production-size batches. One of the most important processes involved in the scale-up of liquid parenteral preparations are mixing. For liquids, mixing can be defined as a transport process that occurs simultaneously in three different scales, during which one substance (solute) achieves a uniform concentration in another substance (solvent). On a large, visible scale, mixing occurs by bulk diffusion, in which the elements are blended by the pumping action of the mixer's impeller. On the microscopic scale, elements that are in proximity are blended by eddy currents, and they create drag, where local velocity and shear–stress differences act on the fluid. On the smallest scale, final blending occurs via molecular diffusion, whose rate is unaffected by the mechanical mixing action. Therefore large-scale mixing depends primarily on flow within the vessel, whereas small-scale mixing is dependent mostly on shear. This approach focuses on large-scale mixing taking into consideration geometric equivalence of vessels, dimensionless numbers corresponding to product/process relationships (such as Reynolds number), as well as scale of agitation levels describing different mixing levels in the vessels.

Table 7.7 lists a nomenclature used in analysis.

Table 7.7 Nomenclature for Scale of Agitation Approach

	Nomenclature
Q	Effective pumping capacity or volumetric flow, in cm^3/s
N	Shaft speed, in s^{-1}
N_{Re}	Impeller Reynolds number, dimensionless
N_Q	Pumping number, dimensionless
D	Diameter of the largest mixer blade, in cm
ρ	Density of the fluid, in g/cm^3
μ	Viscosity of the fluid, in g/(cm/s)
v_b	Bulk fluid velocity, in cm
T	Diameter of the tank, in cm
A	Cross-sectional area of the tank, in cm^2

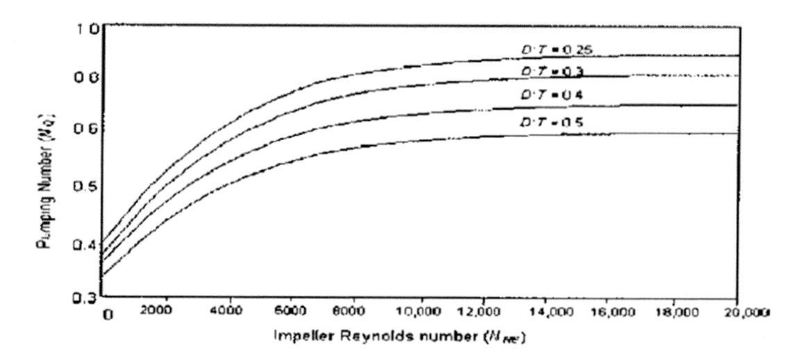

Figure 7.7 Relationship between Pumping number and Reynolds number in turbulent range is close to a straight line; therefore, linear extrapolation is used in analysis to calculate pumping number.

An evaluation of a Reynolds number for a mixing environment using Eq. (7.2) should describe mixing in the vessel. Typically, unless the liquid is a suspension mixing produces a turbulent flow during manufacturing of products, as Reynolds number should be above 2000.

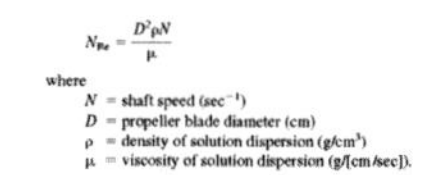

Equation 7.2 Reynolds number equation.

where N is the shaft speed (s^{-1}), D is the propeller blade diameter (cm), ρ is the density of solution dispersion (g/cm^3), and μ is the viscosity of solution dispersion (g/(cm/s)).

It should be noted from Fig. 7.7 that relationship exists between Pumping Number and Reynolds Number in Turbulent Range that is close to a straight line; therefore linear extrapolation is used in analysis to calculate pumping number.

Table 7.8 shows an example calculation used to show equivalence between 2000 L vessel mixing environment and 4000 L vessel environments. Both are using a scale of agitation of approximately 1 (refer to Table 7.9).

Finally, Table 7.9 describes a Scale of Agitation mixing environments' requirements.

In conclusion of the scale-up process calculations, mixing environment used in the original vessel and scaled-up vessel determined to be

Table 7.8 Scale of Agitation Calculation Procedure

Original Vessel—2000 L	Parameters	Units	Equations
Original tank diameter$_{2000\ L}$	121.12	cm	
Original impeller diameter$_{2000\ L}$	15.24	cm	
Original RPM$_{2000\ L}$	350	RPM	
Original shaft speed$_{2000L}$	5.8	s^{-1}	
Liquid density	1.045	gm/cm^3	
Liquid viscosity	0.5	gm/cm/s	
Cross-sectional area$_{2000\ L}$	11522	cm^2	$A = \frac{\pi T^2}{4} cm^2$
D/T ratio$_{2000\ L}$	0.13		D/T
Reynolds number N_{RE}	2832	>2000	$N_{Re} = \frac{D^2 \rho N}{\mu}$
Pumping number N_Q of 2000 L	0.99		$N_Q = 1.1283 - 1.07118\left(\frac{D}{T}\right)$
Pumping capacity Q$_{2000\ L}$	2	cm^3/s	$Q = N_Q N D^3\ cm^3/s$
Bulk fluid velocity$_{2000\ L}$	2	cm/s	$V_b = \frac{Q}{A} cm/s$
	New Vessel—4000 L		
New tank diameter$_{4000\ L}$	176.53	cm	
New impeller diameter	19.20	cm	
D/T ratio$_{4000\ L}$	0.11		
Cross-sectional area$_{4000\ L}$	24,475	cm^2	
Pumping capacity Q$_{4000L}$	43,577	cm^3/s	
Pumping number N_Q	1.01		
Bulk fluid velocity$_{4000\ L}$	6.1	s^{-1}	
RPM$_{4000\ L}$	365	RPM	

a Scale of Agitation 1 with bulk fluid velocity below 3 cm/s which means that the mixing in the vessel requires a minimum to achieve turbulent flow. Similar or better mixing environment could be easily achieved in a new 4000 L vessel with a bulk fluid velocity of approximately 6 cm/s (400 RPM). Therefore mixing environment in a previously utilized for this process 2000 L vessel is equivalent to the mixing to be used in a new vessel and would allow for the process that would result in manufacturing of consistent.

When Process Control and Scale-up strategies are developed, one is ready for Technology Transfer. Most important concepts of this task are:

- Setting up mechanisms for knowledge transfer between process development and manufacturing.
- Identifying process variables and designing adequate control strategies that may be vulnerable because of the technology transfer.

Table 7.9 Process Requirements: The Set Degree of Agitation for Blending Motion

Scale of Agitation	Bulk Fluid Velocity (cm/s)	Description of Mixing
1	3	Agitation levels 1 and 2 are characteristics of applications requiring minimum fluid velocities to achieve the product result
2	6	Agitators capable of level 2 will: • Blend miscible fluids to uniformity if specific gravity differences are less than 0.1 and if the viscosity of the most viscous is less than 100 times that of the other • Establish complete fluid-batch control • Produce a flat but moving fluid-batch surface
3	9	Agitation levels 3–6 are characteristic of fluid velocities in most chemical (including pharmaceutical) industries agitated batches
4	12	Same as 3
5	15	Same as 3 and 4
6	18	Agitators capable of level 6 will: • Blend miscible fluids to uniformity if specific gravity differences are less than 0.6 and if the viscosity of the most viscous is less than 10,000 times that of the other • Suspend trace solids (< 2%) with settling rates of 2–4 ft/min • Produce surface rippling at lower viscosities
7	21	Agitation levels 7–10 are characteristic of applications requiring high fluid velocities for process result, such as mixing of the high-viscosity suspension preparations
8	24	Same as 7
9	27	Same as 7 and 8
10	30	Agitators capable of level 10 will: • Blend miscible fluids to uniformity if specific gravity differences are less than 1.0 and if the viscosity of the most viscous is less than 100,000 times that of the other • Suspend trace solids (< 2%) with settling rates of 4–6 ft/min • Provide surging surface at low viscosities

That is why one must return to Risk Management exercise to assess correlation between information obtained from small-scale to large-scale batches. In addition, this exercise should identify possible differences in scale which could affect process performance, assure process control in place to address this variation, use information from similar existing processes.

Accurate and complete transfer of information:

- defines key terms;
- provides a consistent interpretation;
- allows flexibility for innovative approaches; and
- covers various scenarios.

In Technology Transfer the Key Time limiting Factor is Stability. The studies are usually conducted with product made at the launch or manufacturing site. Amount of data needed and timing for filing the change are dependent on the classification of the drug substance and the complexity of the dosage form. Complex forms or high-risk products such as vaccines, cell therapies products need at least 3 months at filing from three batches. Moderate level forms need 3 months from one batch submitted during the review cycle depends on the content of the original filing (solutions, suspensions) and other dosage forms or minor level require simply a commitment for long-term and accelerated stability at filing. Alternatively the validation batches may be used to confirm the site is under control. The Certificate of Analysis is filed for the batches they are then placed on regular stability and the data are supplied as an annual review.

This should conclude major points to consider for Stage 1 Process Development for Parenteral Products.

In addition to abovementioned points we will include an example of a DoEs for Vaccine product. The Fermentation unit operation will be used in this example.

Stage 1: Design of Experiments Case Study

One of the most important tasks of the Process Validation Stage 1 Process Design is to incorporate QbD into Development Program. Typically one would design, perform and analyze Designs of Experiments based on the Quality Risk Management exercises that would identify Critical Process Parameters and CQAs that they control. DoEs are used to investigate process robustness in order to develop design space control strategy.

A product—subject of this case study is a hypothetical vaccine called ABCtiter. The studies are performed prior to PPQ.

The Quality Risk Assessment Report recommended to perform studies on Fermentation and Purification unit operations. The half Fraction Factorial DoEs was recommended to study the impact of Fermentation and Purification key process parameters (KPP) on CQA.

- **Fractional Factorial Design**
- Factors: 3 Base Design: 3, 4 Resolution: III
- Runs: 6 Replicates: 1 Fraction: 1/2
- Blocks: 1 Center pts (total): 2
- * NOTE * Some main effects are confounded with two-way interactions.
- Design Generators: C = AB
- Alias Structure
- I + ABC
- A + BC
- B + AC
- C + AB
- Design Table (randomized)
- Run A B C
- 1 0 0 0
- 2 + - -
- 3 + + +
- 4 - - +
- 5 - + -
- 6 0 0 0

Figure 7.8 Stage 1: Process development design of experiments case study.

Fig. 7.8 shows a study design by a statistical software.[1] It shows that the study will investigate three factors (CPPs) and will consist of six randomly produced purification batches. To produce six purification batches three precursor fermentation batches would have to be produced.

Table 7.10 lists DoE batches, factors, conditions, and sequence in which they will be produced.

Even though these are small batches entire study takes approximately 8 weeks, as the process is considerably complex, especially Fermentation part. Upon completion of the study process and product testing results are tabulated for Residual Protein (%) after each Elution, Endotoxin (EU/μg) after each Elution, as well as Pentose Yield for Fermentation, after each Elution and Purification. Table 7.11 shows the tabulated results of the Quality Attributes that were studied.

[1] Minitab14 is used for this study.

Table 7.10 Design of Experiments Factors Summary

Run Order	Fermentation		Purification		
	Fermentation Duration (h)	Fermentation Batches	Purification: Buffer Temperature (°C)	Elution: Times (× 1 Target)	Actual Elution Times
Fermentation 1	2 h after reaching Stationary Phase[a]	First Fermentation Batch	30	1.25	CTAB[b]; Elution (1–4); Time: 75 min
					65% EtOH[c]; Elution (1–3); Time: 75 min
					65% EtOH; Elution (4); Time: 45 min
Purification 2	2 h after reaching Stationary Phase[a]	First Fermentation Batch	30	1.25	CTAB; Elution (1–4); Time: 75 min
					65% EtOH; Elution (1–3); Time: 75 min
					65% EtOH; Elution (4); Time: 45 min
Fermentation 3	4 h after reaching Stationary Phase[a]	Second Fermentation Batch	20	1	CTAB; Elution (1–4); Time: 60 min
					65% EtOH; Elution (1–3); Time: 60 min
					65% EtOH Elution (4) Time: 30 min
Purification 4	4 h after reaching Stationary Phase[a]	Second Fermentation Batch	40	1.5	CTAB; Elution (1–4); Time: 90 min
					65% EtOH; Elution (1–3); Time: 90 min
					65% EtOH; Elution (4); Time: 60 min
Fermentation 5	Stationary Phase[a]	Third Fermentation Batch	20	1.5	CTAB; Elution (1–4); Time: 90 min
					65% EtOH; Elution (1–3); Time: 90 min
					65% EtOH; Elution (4); Time: 60 min
Purification 6	Stationary Phase[a]	Third Fermentation Batch	40	1	CTAB Elution; (1–4); Time: 60 min
					65% EtOH; Elution (1–3); Time: 60 min
					65% EtOH; Elution (4); Time: 30 min

[a] *Stationary Phase is defined as two consecutive drops or leveling off of OD as determined by hourly OD_{600} measurements after reaching peak OD. (OD is an optical density of the culture. It estimates the culture growth and metabolic activity of the cells. OD is a logarithmic function. An increasing the number of light absorption unit by one corresponds to diminishing by 10 of the intensity of light passing through the sample.)*

[b] *CTAB is a cetyl trimethylammonium bromide, a cationic detergent, facilitates the separation of polysaccharides during purification while additives.*

[c] *EtOH is Ethyl Alcohol.*

Table 7.11 Critical Responses Summary

Purification Run Number	Fermentation Time	Purification Buffer Temperature (°C)	Purification Elution Time (x Target Elution Time)	Fermentation Pentose Yield (gm)	Elution Cycle #1 Pentose Yield (gm)	Elution Cycle #1 Residual Protein (%)	Elution Cycle #1 Endotoxin (EU/μg)	Elution Cycle #2 Pentose Yield (gm)	Elution Cycle #2 Residual Protein (%)	Elution Cycle #2 Endotoxin (EU/μg)	Elution Cycle #3 Pentose Yield (gm)	Elution Cycle #3 Residual Protein (%)	Elution Cycle #4 Residual Protein (%)	Elution Cycle #4 Endotoxin (EU/μg)	Purification Pentose Yield (gm)
1A	2 h[a]	30	1.25	8.24	3.17	2.12	943	2.82	0.36	112	2.53	0.36	0.4	0.494	1.83
1B	2 h[a]	30	1.25	8.24	3.07	1.70	999	2.57	0.39	142	2.18	0.46	0.55	0.165	1.95
2C	4 h[a]	20	1.00	7.34	2.99	1.27	910	2.72	0.51	34	2.12	0.64	0.69	0.154	1.96
2D	4 h[a]	40	1.50	7.34	2.81	1.12	794	2.48	0.52	21	2.17	0.63	0.81	0.935	1.56
3E	At Transition Phase	20	1.50	5.99	2.77	0.93	702	2.39	0.50	31	1.77	0.67	0.89	0.396	0.76
3F	At Transition Phase	40	1.00	5.99	2.63	0.80	466	2.21	0.63	34	1.59	0.84	1.16	0.149	0.94

[a] *Post transition Phase.*

In terms of a DoEs, Table 7.11 summarizes results for experimental runs critical responses: Pentose Yield (gm), Residual Protein (%), Endotoxin (EU/μg Polysaccharide) at specific points in the process (end of fermentation and at the end of the four CTAP/EtOH iterations). We will evaluate some of the examples of analysis performed on these data and summarize some of the conclusions that can be drawn from the study in the next series of charts. Table 7.12 and Figs. 7.9–7.20 illustrate the analyses performed during evaluation of the results from this study.

Upon completion of evaluation of the results from this study the following major observations were noted and will be used while drafting Control Strategy for ABCtiter vaccine:

- The 2-hour post Transition Phase fermentation provided a balance between material production and minimization of cellular lysis which will cause the production material higher resolubilization difficulty. Although this may result in a slightly higher Endotoxin and Residual Protein results.
- The shortest (reaching Transition Phase) fermentation time results in the lowest yield due to the absence of cellular debris. As the cells enter stationary phase they begin to lyse. Concomitant with the production of cellular debris, release of production material is enhanced. Yields for these batches are only mildly affected by buffer temperature and/or duration of elution.
- The longest (4-hour post Transition phase) fermentation results in a slightly lower production material Yield than 2-hour post Transition Phase fermentation but with higher buffer temperature and longer elution time will result in lowest Endotoxin and reduced Protein results.
- Finally it was observed that after each elution there was a log (× 10) reduction in Endotoxin results (EU/μg material) as was historically observed during manufacturing of similar formulated vaccines.

The recommendations from these observations included:

- A fermentation time post stationary phase should not be longer than 3 hours.
- An elution time could be 1.25–1.5 times target (or after second iteration)

Table 7.12 Design of Experiments Factors Versus Responses Interpretation Summary

Study Factors	Factors Versus Responses	Figure Numbers	Interpretation of Results
Pentose Yield Results	Fermentation Duration (Post Transition Phase) versus Yield at Specific Process Points	9	Pentose yield results plotted against fermentation time (at specific points in the process, the end of the four CTAP/EtOH iterations and at the end of purification) shows medium to very strong [$R^2 = 0.69$ (69%) to 0.97 (97%)] correlations using polynomial regression. It is evident that the highest yields at every point have been attained at or just after 2-h post Transition Phase
	Elution Time versus Yield	10	Pentose yield plotted against elution time (at specific points in the process, the end of the four CTAP/EtOH iterations and at the end of purification) shows some level of correlation [$R^2 = 0.33$ (33%) to 0.53% (53%)] using polynomial regression
			The highest yields at every point seem to correlate to the elution times above 1.25 times the target elution times.
	Buffer Temperature versus Yield	11	Pentose yield plotted against buffer temperature (at specific points in the process the end of the four CTAP/EtOH iterations and at the end of purification) show some correlations [$R^2 = 0.34$ (34%) to 0.52% (52%)] using polynomial regression
			The highest yields at every point seems to correlate to the buffers that have been heated temperatures above 30°C
	Fermentation Time versus Yield Optimization	12	This chart is plotted using polynomial equations attained after fitting purification Pentose Yield data versus fermentation time posts Transition Phase. This chart is a simulation based on the best fit equations
			It is evident that the best Pentose Yields results are predicted for processes that extend fermentation time 2 and 3 h post Transition Phase
Residual Protein Results	Residual Protein Main Effects Plots for Each Elution	13	Residual Protein Main Effects Plots show that higher buffer temperature and longer elution time result in lower residual protein results. This is particularly apparent when fermentations with times above 2 h post Transition Phase where used, especially after second elution
	Fermentation Time, Buffer Temperature and Elution Time versus Residual Protein	14, 15, 16	Fermentation Time, Buffer Temperature and Elution Time were plotted against Residual Protein. These plots show that residual protein may be reduced in lots produced with fermentations that were held more than 2 h beyond Transition Phase by increasing buffer temperature and prolonging elution times
Endotoxin Results	Endotoxin Interaction Plots for Each Elution	17	Endotoxin Interaction Plots show close to interaction results between higher Buffer temperature and longer elution duration (especially after Elution 3) that correspond to lower Endotoxin results
	Fermentation Time, Buffer Temperature and Elution Time versus Endotoxin	18, 19, 20	Fermentation Time, Buffer Temperature and Elution Time were plotted against Endotoxin. These plots show that there is a log ($\times 10$) reduction in Endotoxin that is noted after each elution. In addition, it is shown that endotoxin may be reduced in lots produced with fermentations that were held more than 2 h beyond Transition Phase by increasing buffer temperature and prolonging elution times.

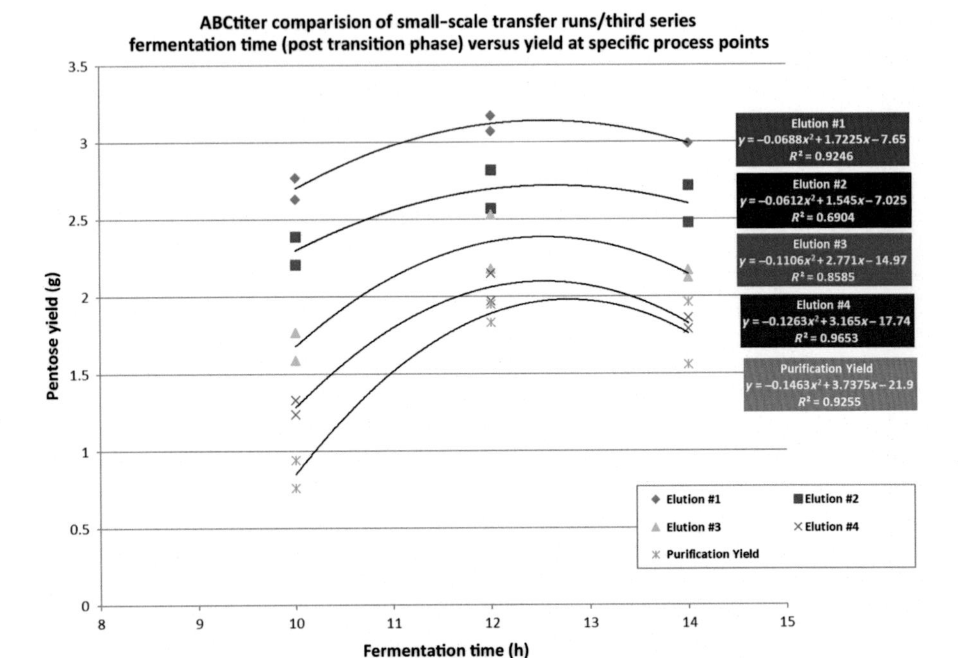

Figure 7.9 Fermentation duration (post transition phase) versus yield at specific process points.

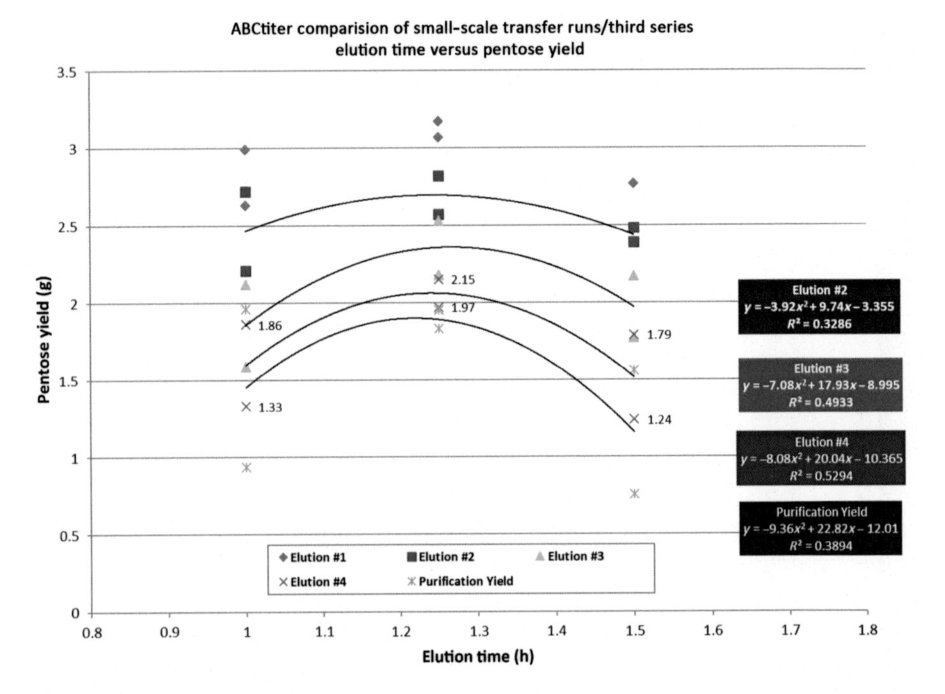

Figure 7.10 Elution time versus pentose yield.

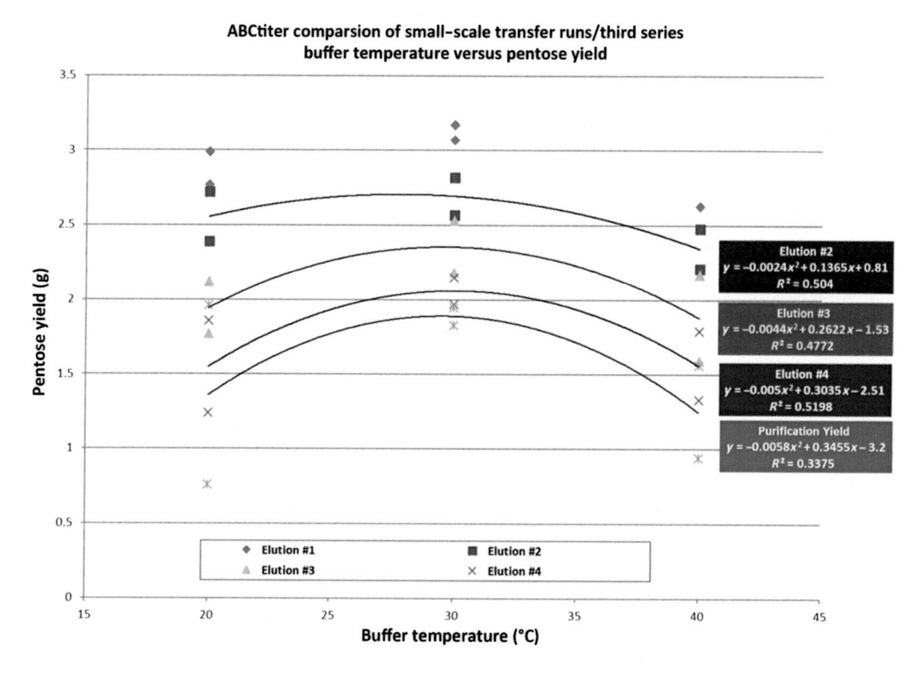

Figure 7.11 Buffer temperature versus pentose yield.

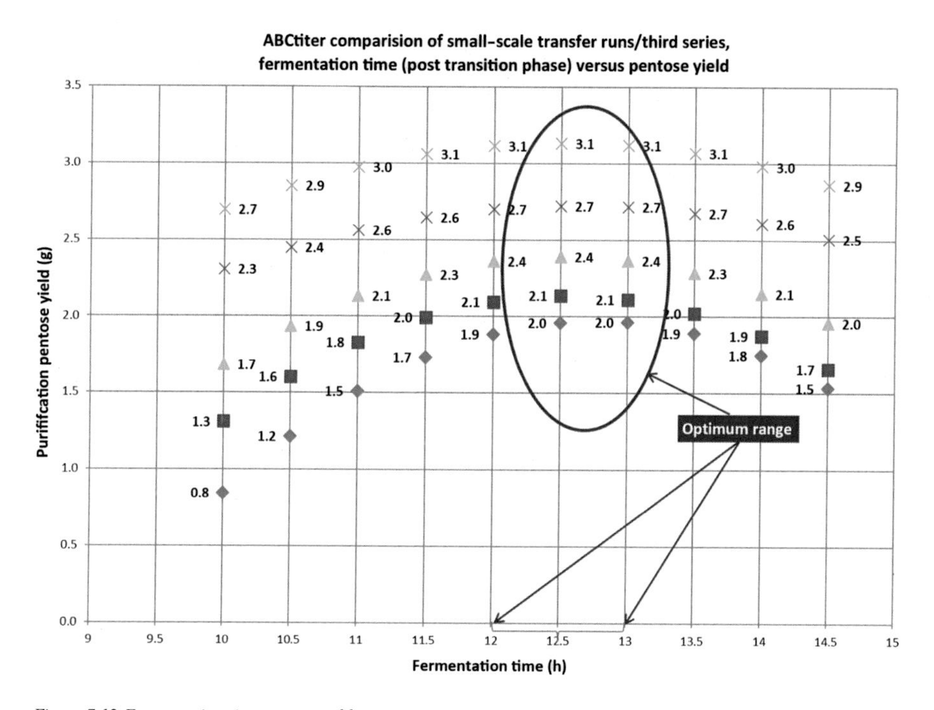

Figure 7.12 Fermentation time versus yield optimization.

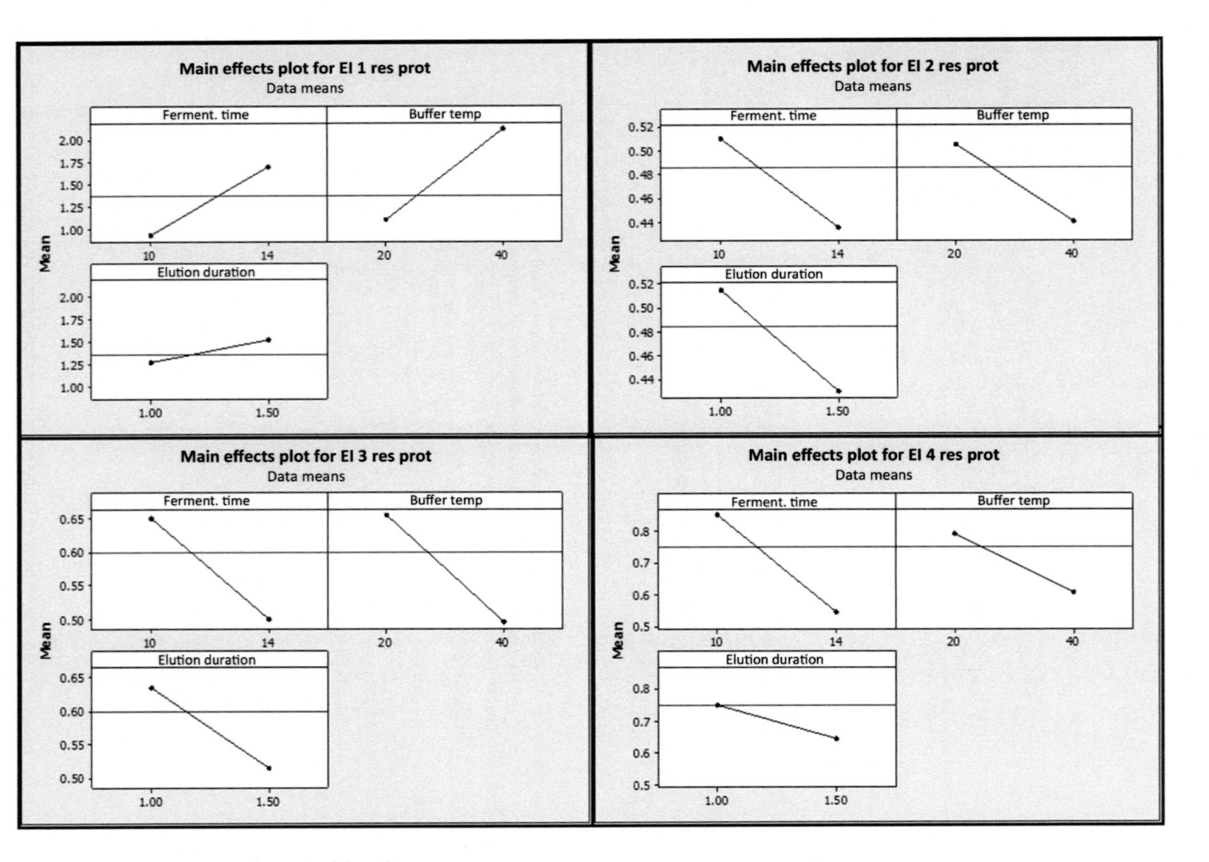

Figure 7.13 Residual protein main effects plots for each of four elutions.

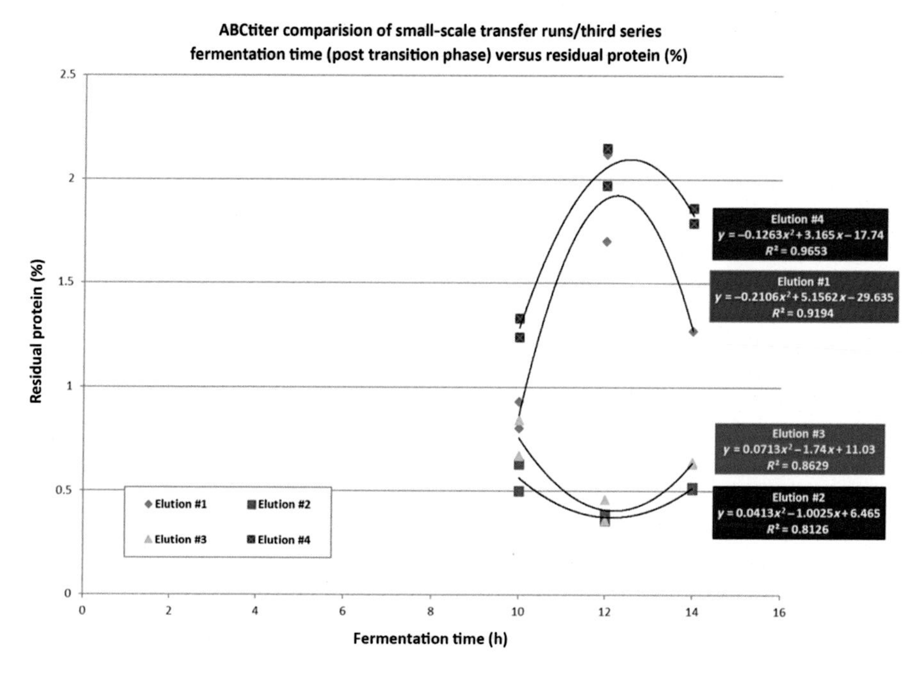

Figure 7.14 Fermentation time versus residual protein.

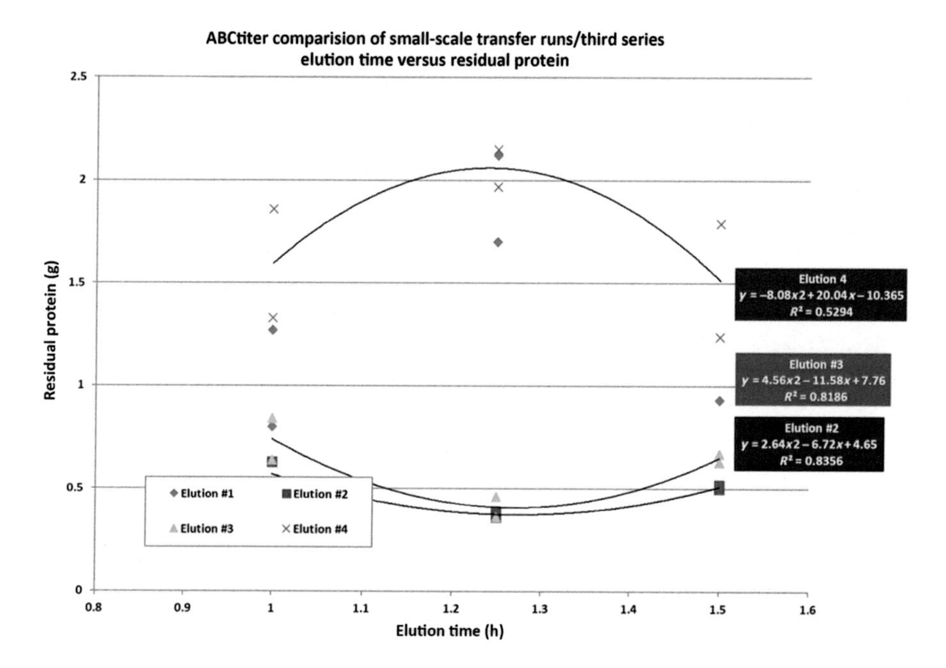

Figure 7.15 Elution time versus residual protein.

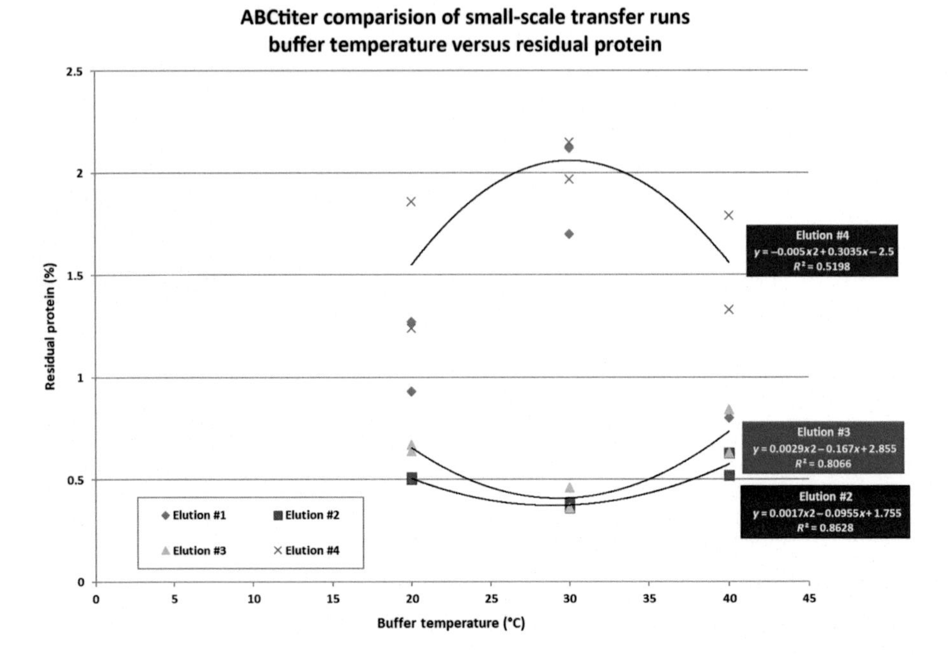

Figure 7.16 Buffer temperature versus residual protein.

- A buffer temperature could be 35°C–40°C (after second iteration), as the greatest effect is seen at the second iteration while the effect is less prominent for the other iterations.

This study is paving way for scientifically established controls that result in higher yields and lower contaminants, which will ultimately result in higher quality build into the process.

Final Notes About Stage 1 Process Design

We have tried to discuss only major concepts of Stage 1 including usage of Risk Management, Quality-by-Design and other scientific tools that should help in establishing effective and efficient Process Design and Development program. The final recommendations include a notion that Research & Development personnel should be working hand-in-hand with Technical Services, Operations, Quality Assurance, and Quality Control personnel while periodically updating upper management at significant substages of development. This stage of Process Validation for Parenteral product is probably the most significant in

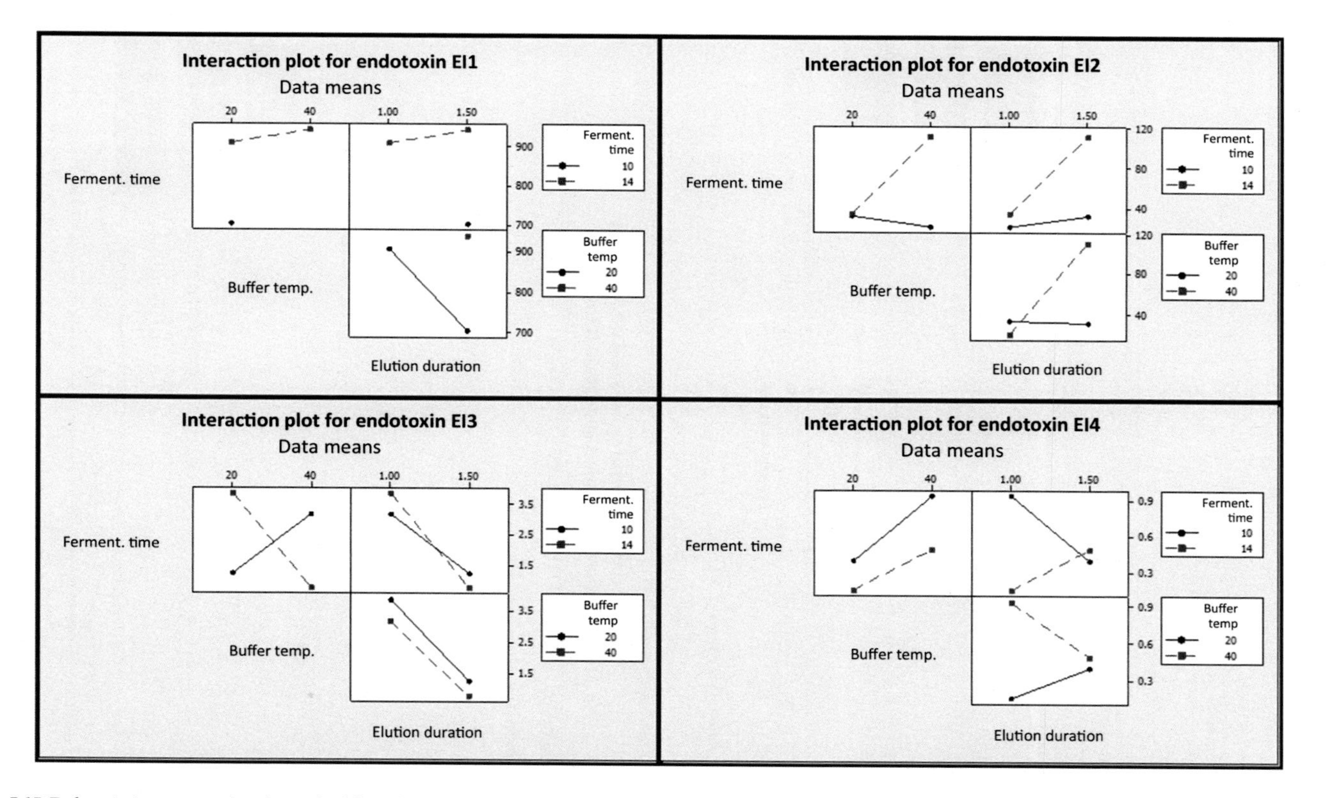

Figure 7.17 Endotoxin interaction plots for each of four elutions.

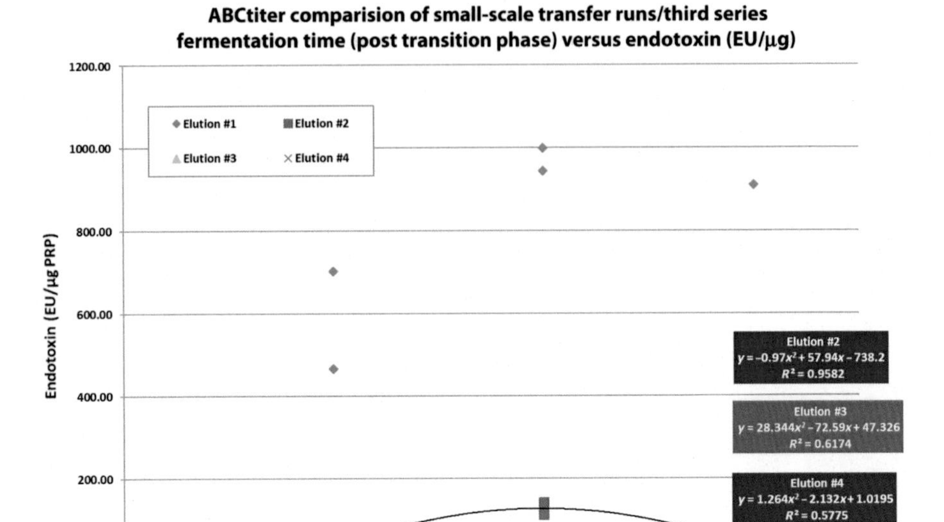

Figure 7.18 Fermentation time versus endotoxin.

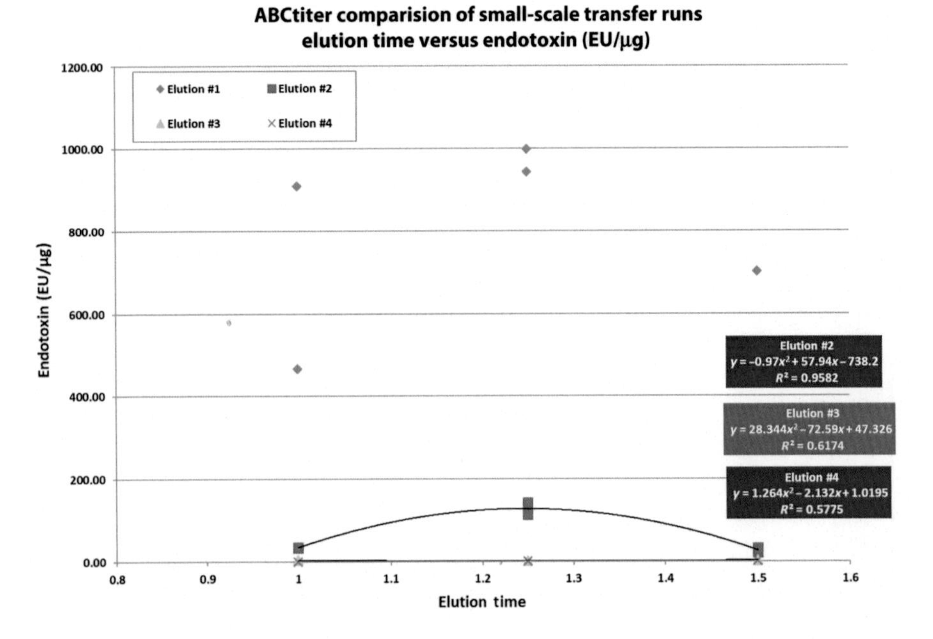

Figure 7.19 Elution time versus endotoxin.

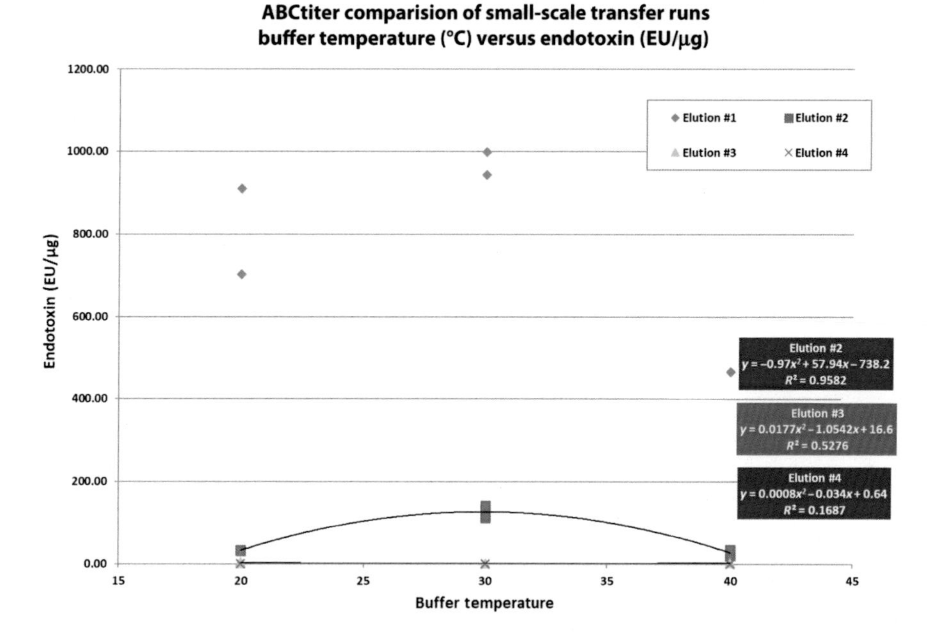

Figure 7.20 Buffer temperature versus endotoxin.

an entire life cycle of a product and a process and therefore requires almost attention, as it becomes a pillar on which process will reside for the rest of its life.

References

FDA, 2011. Process Validation Guidance: Principals and Practices.

FDA, 2004. Report Pharmaceutical CGMPs for the 21st Century—A Risk-Based Approach, Final Report, Department of Health and Human Services, US Food and Drug Administration, September.

ICH, 2009. Pharmaceutical Development, Q8(R2), Step 4 version, August, International Conference for Harmonization.

Parenteral Drug, 2011. Scale-up by I. Gorsky in Pharmaceutical Process Scale-Up Handbook, Drugs and The Pharmaceutical Sciences Series Volume 118, Marcel Dekker, NY, Parenteral Drug Scale-Up Chapter, First Edition, 2001, Second Edition, 2006, Third Edition, 2011.

PDA, 2012. Technical Report 60, Process Validation – Risk Based Lifecycle Approach.

CHAPTER 8

Process Validation Stage 2: Parenteral Process Performance Qualification

Harold S. Baseman[1] and Igor Gorsky[2]

[1]Chief Operating Officer, Valsource Inc, Jupiter, FL, United States, [2]Senior Consultant, ConcordiaValsource LLC., Downingtown, PA, United States

This chapter will address the second stage of the process validation life cycle, process qualification. It follows Process Design and is followed by Continued Process Verification (refer illustration to Figure 1.1 of Chapter 1: Process Validation: Design and Planning). Aseptic process simulations or media fills are part of this stage and was discussed in detail in Chapter 1, Process Validation: Design and Planning. This chapter will focus on aseptic processing, although terminal sterilization may also be used for some parenteral products.

General Principles of Stage 2 Life Cycle Approach to Process Validation for Parenteral Products

Process validation is a matter of obtaining confidence that a process is capable consistently performing to a level that will yield product of a prescribed level of quality. In this way, assurance that the product manufactured with the process will meet quality specifications is provided. While testing of parenteral products can provide some level of quality assurance, testing alone will not provide desired confidence that all the product will meet quality specifications.

As discussed earlier, in Chapter 1, Process Validation: Design and Planning, of this book, this was notably illustrated in 1970–71 when outbreaks of *Enterobacter cloacae* and *Erwinia* contamination in large volume parenteral bottles resulted in injury and death to patients using IV solutions. The contamination was apparently caused by moisture seeping into a space under the screw cap of the bottles during cooling after sterilization. The caps and stoppers were removed prior to sterility testing, so the contamination was

Principles of Parenteral Solution Validation. DOI: https://doi.org/10.1016/B978-0-12-809412-9.00007-1

not found. This unfortunate example shows why product testing and process monitoring are not enough to assure the quality of the product (Maki et al., 1976).

Out of this and other incidents came the realization that product testing alone would not be sufficient to provide confidence of product sterility. To achieve this objective, one must also prove process capability. This is the foundation of modern process validation.

The US FDA GMPs, part 21 CFR 211.100 requires that: "There shall be written procedures for product and process control designed to assure that drug products have identity, strength, quality, and purity they purport or are represented to possess" (FDA, 2011).

Assurance of the quality of a product can be confirmed by direct observation, testing or inspection of that product (refer *to* Figure 1.1 of Chapter 1: Process Validation: Design and Planning). However that is not always possible. The testing of a sample of a parenteral product for potency or sterility, at this time, would be destructive to the product and render the product useless. Where complete observation is not possible, assurance becomes a matter of prediction that products will have a desired quality. Validation thus would be the prediction of a direct condition, in this case potency or sterility, one cannot fully observe, based on indirect conditions one can observe. These indirect conditions become surrogates that can be observed during process validation testing. And the correlation between these observations and product quality attributes define the acceptance criteria for the tests and studies (Fig. 8.1).

Recognizing this, the US FDA published Guidance on Process Validation defined process validation in its May 1987 Guideline on General Principles of Process Validation *as establishing documented evidence which provides a high degree of assurance that a specific*

Proving assurance is a balance of ...
Observation and Prediction
Validation is prediction of outcome that cannot be fully observed, based on analysis of conditions which can be observed.

Inspection
Prediction

Figure 8.1 Validation definition.

process will consistently produce a product meeting its predetermined specifications and quality characteristics. The guidance restated the regulatory requirements for validation, the link to product quality assurance. It introduced such concepts as equipment qualification, installation qualification, prospective validation. It also linked product testing, at an enhanced level to process performance. This resulted in the adoption of replicate process batches as the criteria for process acceptability.

However the use of replicate batches as the primary determinant of process capability, alone, is not sufficient to identify or confirm the control of all significant process variables, and therefore may not be not an adequate validation approach. The practice of assuring and demonstrating process quality and reliability by successfully running multiple, traditionally three process batches may not consider all process variables. This is evidenced by process failures that may occur after processes are validated by this approach. In other words, if the process is validated to a high degree of assurance by successfully running replicate batches, then process failures should be rare. However these failures do occur. The reason for failures may be a result of raw material changes, human error, changing environmental, or process conditions, and so on. All these can be considered as unforeseen or inadequately addressed process variables.

Recognizing this issue, the FDA revised the guidance in January 2011, redefining and clarifying the definition for process validation as *the collection and evaluation of data, from the process design stage throughout production, which establishes scientific evidence that a process is capable of consistently delivering quality products* (FDA, 2011).

The revised guidance presented an approach to process validation based on continuous learning and process understanding through a process's life cycle. It involves knowledge acquisition, process challenge, monitoring, and results evaluation. Process validation *establishes and demonstrates process control* the life cycle concept links product and process development, qualification of the *commercial manufacturing process*, and maintenance of the process in a state of control during routine commercial production (FDA, 2011).

The 1987 and 2011 FDA process validation guidance definitions appear similar, but there are some interesting differences, that may signal an evolution of the approach over the span of nearly 25 years.

In 1987, the definition emphasized establishing documented evidence, while in the 2011 version, the emphasizes are on establishing scientific evidences. What is the difference between documented evidence and scientific evidence? In part, the difference is that the evidence referred to in the latter revision must have a scientific and logical basis. Scientific basis means that the evidence must be logical and defendable, consistent with statistical and scientific principles.

The 2011 FDA Process Validation Guidance for Industry proposed three process life cycle stages for validating a process. These include Stage 1: Process design, Stage 2: Process qualification, and Stage 3: Continued process verification (refer to Figure 1.3 of Chapter 1: Process Validation: Design and Planning). Process design establishes the process variables and control strategy and measures needed to ensure that those variables do not adversely affect the performance of the process to the extent that the process will fail. As such it is the stage that establishes confidence in the process. Process qualification further tests those control measures against a prescribed set of acceptance criteria. As such, it provides an opportunity to ensure that no significant variables have been missed or inadequately addressed. Continued process verification allows for a continuous and ongoing evaluation of the process. As such it provides the opportunity to detect and recognize variables that may appear in the process. The addressing of those variables, by definition, improve the process. European and other global guidance emphasize similar life cycle, process understanding-based approaches (FDA, 2011; EU Guidelines, 2015).

While the FDA guidance does not specifically refer to aseptic processes, the principles presented in the guidance are logical and can be used as an approach for validating such processes. Aseptic processing does, however, have some unique aspects and considerations.

In September of the 2008, the US FDA announced revisions to the US GMPs part 211. Among the changes was the insertion the word "aseptic" into Section 211.113.

211.113 Control of microbiological contamination.

(b) Appropriate written procedures, designed to prevent microbiological contamination of drug products purporting to be sterile, shall be established and followed. Such procedures shall include validation of all *aseptic* and sterilization processes.

Further, the preamble to the Federal Registry stated "... industry routinely conducted validation studies that substituted microbiological media for the actual product to demonstrate that its aseptic processes were validated. These parts of validation studies are often referred to as media fills. ... this revision clarifies existing practices and serves to harmonize the CGMP requirements with Annex 1 of the EU GMPs, which requires such validation." Some in the industry, including members of the PDA commenting committee indicated concern over the change in the context of the preamble statement. The concern was that the preamble statement appeared to indicate that the running of media fills was the sole method for validating the aseptic process. And in fact, successfully completing a series of media fills, three, would validate the process (PDA, 2008a).

Some on the committee, including this author, were concerned that one could not validate and aseptic process using the same approaches as validating a sterilization process. There are a few reasons for the difficulty in validating an aseptic process.

First, there is the variability inherent in the process. Most aseptic processes involve human manipulations, activities, and interventions. Humans performance is highly variable. Even semiautomated processes in relatively closed isolators involve interventions, where gloves are not sterile, but decontaminated can pose a risk to product contamination. Aside from Isolators, most aseptic processes are somewhat open to the environment. The behavior of microorganisms, as demonstrated by the correlation of environmental counts to product contamination, is also relatively unpredictable.

Second, it is difficult to detect actual contamination in the environment or product on a real-time basis. Therefore correlating what can be measured with the desired or undesired outcome is difficult. Validation as defined earlier is the prediction of an outcome that cannot be fully observed based on the evaluation of conditions and metrics that cannot be observed.

Third, aseptic processing involves the proving of a negative, the absence of an event, rather than an action. To illustrate, consider the comparison of the two processes mentioned in cfr 211.113b, sterilization and aseptic processing. In sterilization, a level of contamination is assumed, and steps or actions are taken to destroy or eliminate that contamination. Because it is known how much energy is required to destroy a given level of microbiological contamination, there is a strong correlation between what can be observed and the desired outcome. Proving that outcome then becomes a matter of proving that the action, application of moist heat at a given temperature for a given period under a given pressure, has been applied to transfer there required energy and destroy the microorganisms. However, in the case of aseptic processing, we are not taking actions to eliminate contamination, instead we are taking actions to prevent an event, the contamination of product. Proving that the event has not happened is more difficult that proving that the transfer lethal energy action has occurred.

Because of the complexities and challenges of aseptic processes, it useful consider a multistage, holistic approach to validate the aseptic process and to provide this level of proof. This is consistent with the FDA Life Cycle approach recommended in the FDA and other process validation guidance. It is important to understand the relative effectiveness and limitations of building process confidence at each stage the process life cycle and the value of process design rather than product testing to ensure product quality. Contamination control and aseptic process control are a series of actions. One can obtain confidence that these actions are effective through process design, qualification, testing, and monitoring, each of which aligns with the stages of life cycle process validation.

Process Design—The most effective way to ensure process performance and contamination control is to design the process correctly, considering and addressing sources of contamination and variability. If the process works, then it works. No amount of qualification and testing can compensate for a process that does not work.

Process Qualification—Qualification and validation will help identify process variables and sources of contamination wrath have been missed in process design. In addition, qualification and validation are a means to demonstrate to decision makers, auditors, investigators, and regulators that the process has been adequately and effectively controlled. As such qualification and validation serve as an important function. However, as stated earlier, good qualification and validation still does not compensate for inadequate or poor process design.

Ongoing or Continued Process Verification—Testing and Monitoring.

Testing—Incoming, in-process, end of process, or finished product testing can provide additional confidence that the process is still performing as designed. This testing may include component inspection, product inspection, filter integrity testing, pre-use post sterilization integrity testing, container closure testing, sterility testing, and so on. However, in most cases, this testing is destructive and can only be performed on a limited number of samples. While, product and attribute testing can be effective in presenting the quality of the attribute for that sample, it may not be a good indicator of process capability and at best it may be a lagging indicator of process performance.

Monitoring—Monitoring involves viewing and accessing the process in real time to evaluate performance. Monitoring has two benefits. (1) It can be used to determine the conditions by which the process is performing at the time of monitoring. (2) It can be used to predict the future performance of the process. To be effective, monitoring of the process must correlate to the condition or quality of the product. In aseptic processing, this may be difficult. However where the correlation is sound and well understood, monitoring can be an indicator of process performance, if process variables are still adequately controlled, and whether the process has remained in a qualified state.

The Line of Sight Approach to Process Understanding and Process Validation

Stage 1 of the life cycle approach to process validation, process design is where we determine what process steps, conditions, and controls are needed to design the process to meet the requirements of the process. It begins with a recognition of product definition, as presented in the critical attributes of the product. These are the aseptic of the product that define its usefulness, its quality if you will. Quality can be defined in a series of critical quality attributes (CQAs), without any one of which the product is no longer acceptable or useful. These attributes include elements resulting in product safety, purity, identify, and efficacy. Once the CQAs are understood, one can determine the process steps and conditions needed to establish and maintain those CQAs. It is then necessary to identify process variables. These are the elements that may change and change to the extent that the CQAs may be adversely affected. Once these are identified, it is prudent to put in place process controls to mitigate the risk of process variables adversely affecting the CQAs. The process design is complete when the process steps, conditions and controls are in place.

These process controls are what must then be challenged and tested during Stage 2 or process qualification phase. This chapter will present the principles of Stage 2, process qualification, in relation to parenteral processes and explore the reasons why these principles should be in place. However it is important to restate that Stage 2 must be built on a strong Stage 1, process design platform. Otherwise, the process qualification studies and tests will not likely align with be effective in meeting any real objective.

As shown in Figure 1.2 of Chapter 1, Process Validation: Design and Planning, the line of sight approach to process validation begins with determining the criticality of quality attributes, then determining how the process steps and respective controls effect those attributes. The testing that is conducted in Stage 2 should link to the CQAs. This means every test function should be linked back to a CQA and every CQA should be addressed by a test function (Long, 2013).

A risk assessment can help determine the effect that process steps and critical process parameters (CPPs) have on CQAs and the variables associated with these steps (refer to Register shown in Table 8.1).

Stage 2: Process Qualification

The US FDA defines Stage 2 as the qualification of the process. The process design, as determined and developed in Stage 1, is the place where one gains confidence that the process works. Stage 2, process qualification, is the place where one confirms the effectiveness of the control strategy designed to provide process performance. One should not wait or rely on process qualification to prove performance of the process. If one does so, then there is a risk that the process performance will not be satisfactory. If so, it is usually too late to easily change the process. In most cases, companies choose to live with or find ways to work around insufficient processes. However even the most well-designed processes may have areas of variation, weaknesses

Table 8.1 An Example of CQA, Variables and Control Strategy Register

CQA	Process Step/ Condition	Variable	Control Strategy	Test
Sterility	Filler parts sterilization	Parts do not reach lethal temperature for requisite time	Continuous monitoring of autoclave temperature and time, as recorded on a chart recorder	Temperature mapping and heat penetration studies
Sterility	Filler parts sterilization	Sterilized parts are contaminated during transfer and storage	Parts are wrapped in microbial barrier material and stored for a finite period under clean conditions	Package design qualification and wrapped parts microbial intrusion challenge
Potency	Mixing of ingredients during formulation	Product is not mixed for enough time	Procedures are set, and compounders trained to maintain agitation at a prescribed seed for a set period	Mixer qualification, product homogeneity sampling study
Purity	Cleaning of compounding vessel	Rinse does not reach all product contact surfaces	Install spray ball to apply rinse water to all surfaces	Cleaning validation with rinse and swab sampling
Identity	Labeling of finished product	Incorrect lot number is placed on label	Use automated reader to check all labels for correct coding	Reader and system qualification with statistical sampling
Safety	Inspection for particulates	Inspectors are not able to see smaller particles	Install suitable lighting, set optimal conveyor speeds and rotate inspectors so that they do not become fatigued	Challenge with representative defects

that were not apparent during process design, appeared after process design, or were not adequately addressed during process design.

Stage 2 provides a final opportunity to catch an unaddressed or underaddressed process variation or weakness prior to committing the process to commercial manufacturing. As such, it must build not the information obtained and developed through process design. Every qualification test should be able to be linked back to the proof that the control strategy designed to establish or protect a given CQA is performing properly. And each CQA control strategy should be represented by one or more qualification tests.

The US FDA Process Validation Guidance Divides Stage 2 Process Qualification into two parts. The first part is sometimes referred to as Stage 2a addresses the qualification of equipment, facility, and utilities to demonstrate they are suitable for their intended use and perform properly. The objective is to eliminate or minimize equipment, facilities, and utilities as a process variable. As stated in the guidance: "Proper design of a manufacturing facility is required under part 211, subpart C, of the CGMP regulations on Buildings and Facilities. It is essential that activities performed to assure proper facility design and commissioning precede PPQ. Here, the term qualification refers to activities undertaken to demonstrate that utilities and equipment are suitable for their intended use and perform properly. These activities necessarily precede manufacturing products at the commercial scale" (FDA, 2011).

Stage 2a will test and demonstrate that the equipment, facility, instrumentation, and systems designed to establish, protect, and control the process, do so in a reliable, predicable, and adequate manner. It is a formal process with written procedures, protocols, and prescribed acceptance criteria. Qualification of a given system is performed after that system's operation and settings required for the process have been established. Qualification, when performed properly, reduces the variability of system performance as a root cause of process failure. Once qualification has been completed, any changes to the system or settings must be addressed and approved through a formal change control procedure that may require repeat of one or more of the qualification test functions. Periodic assessment of the system performance is recommended to ensure that the system is maintained in a qualified state and is performing to a level consistent with process requirements.

The objective of Stage 2 is to determine if the process is capable of reproducible commercial manufacturing (FDA, 2011). To assess the capability of the process, one must first understand the reliability of the mechanical systems, the facility, equipment, utilities, and instrumentation that support the process. If the performance of the systems is not reliable, in other words, if the system performance is a variable, then it may not be possible to determine the capability of the process. It will be difficult to determine whether process failure is a result of inadequate process or system failure. Once Stage 2a has been completed and the reliability of equipment, and supporting utilities the systems has been established, it is possible to move to Stage 2b—the process performance qualification or PPQ. PPQ will establish confidence that the process is capable of provide the CPPs and conditions needed to meet or maintain CQAs of the product. PPQ will also provide another opportunity to test and confirm that process system variables have not been missed or inadequately addressed.

All qualification activities should be performed using a written, formal, quality-approved protocol with written, formal, quality-approved standard operating procedures (SOPs). Test and study acceptance criteria should be prescribed prior to the performance of the qualification tests and studies. Documentation should follow good documentation practices. Deviations from the protocol or discrepancies found during the performance of the qualification should be addressed by a formal SOP. Changes to SOPs discovered or determined during the performance of qualification should follow a change control procedure and process.

The following of some of the common steps to accomplish the objective of Stage 2a (refer to Figure 1.4 of Chapter 1: Process Validation: Design and Planning).

1. *DQ or Design Qualification*—The objective of DQ is to confirm that the system design meets user requirement specification, process requirements, and is compliant with relevant GMP expectations. DQ is a review of process and system drawings, submittals, specifications, and other documentation. DQ may also confirm that the system has adequate design features required for qualification testing and sampling. DQ may also be used to provide the user with information needed to prepare operating and maintenance procedures and qualification protocols.
2. *FAT or Factory Acceptance Testing*—The objective of FAT is to test the system at the fabrication site to determine that the system is ready for shipment to the user site. FAT involves work and tests performed by the vendor at the fabrication or factory site confirming that system is suitable for delivery to the user site. Where possible, it is recommended that the FAT be aligned with tests to be performed by the user during commissioning and qualification and specified in procurement requirements or in a quality agreement. Data and information obtained from FAT may be used to support commissioning and qualification, providing the information is accurate, properly document, no material changes are made or if made are not addressed after FAT, and variables associated with packing, handling, and transport are taken into consideration. Training of user operations and maintenance personnel may be conducted during the FAT
3. *SAT or Site Acceptance Testing*—The objective of SAT is to test the system once it is installed at the user site to determine if it has

been installed properly and in proper working order. SAT involves testing the system after it is installed prior to turnover to the user. As is the case with FAT, it is recommended that the SAT be aligned with tests to be performed by the user during commissioning and qualification and specified in procurement requirements or in a quality agreement. Data and information obtained from SAT may be used to support commissioning and qualification, providing the information is accurate, properly document, no material changes are made or if made are not addressed after SAT. Training of user operations and maintenance personnel may be conducted during the SAT

4. *Commissioning*—The objective of commissioning is to ready the system for operation. Commissioning is usually performed prior to qualification, but after the system is installed and in basic working order. The tests performed during commissioning may be like the tests performed during qualification. Commissioning may not completely take the place of qualification, because commissioning still may involve adjustments to the system to optimize performance and meet process requirements. However some of the tests performed during commissioning, if properly documented, may be used as part of the qualification, providing changes to the system have not been implemented between the time of commissioning and the time of qualification. Commissioning may not require the same formality of documentation or approval as qualification. However it is important that the information developed or leveraged during commissioning be reliable (ISPE, 2019).
5. *IQ or Installation Qualification*—The objective of IQ is to confirm that system has been installed according to design. IQ may involve additional document review combined with physical walk down and inspection of the system. IQ may also leverage or include information gathered form tests and activities performed during commissioning, providing the system has not materially changed during or since commissioning or that any changes have been addressed in a formal change control procedure.
6. *OQ or Operational Qualification*—The objective of OQ is to confirm that the system operates to its intended use (or its capability), relevant to the process requirements. Tests are designed to confirm that the system will operate in a reliable and predictable manner. It may not be necessary to test the system to failure or to test to capacity if that capacity significantly exceeds the process

requirements. Operational settings, parameters, and conditions should be set prior to the performance of the operational qualification tests. Data and information may be used from commissioning or other installation-related test results, providing those results are accurate and suitably documented, and no material changes are made after commissioning or if made are not properly addressed using approved change control procedures. Instructions for the operation of the system, including settings should be noted in an approved SOP or protocol.

7. *PQ or Performance Qualification*—The objective of performance qualification is to confirm that the system can perform the process. The system is tested at process settings, under process conditions. The process should meet specifications and acceptance criteria during the tests.

Some Points to Consider for the Qualification of Equipment and Systems

Prior to performing of Operational Qualification and Performance Qualification, all facility, critical utilities, support systems, instrumentation, control and monitoring equipment, and software should be qualified. The following section of the chapter presents some additional points to consider for equipment, facility, utility, system, and instrument qualification, where the aforementioned qualification activities can be applied. These points to consider are not meant to be all inclusive.

Critical Utilities

Critical utilities should be qualified prior to the qualification of the equipment they support. Critical utilities are those that support the parenteral manufacturing or testing process, without which the equipment, instrumentation, and/or process will not reliably result in product meeting CQA specifications. Critical utilities provide services such as compressed air, vacuum, compressed nitrogen, cooling water, clean steam, and electric power. Unless circumstances do not allow, only qualified critical utilities should be used during process equipment and process operational, performance, and process qualifications. Critical utility qualification should consider not only the draw or volume of the utility needed to support a given piece of process equipment, but the full draw or volume of the utility needed to support all equipment and process needs that might be occurring at any one time.

Qualification should address the generation of the utility and the distribution of the utility. The utility should be shown to reliably deliver the support required by the equipment, instrumentation, and/or process. Each critical utility may have its own CQA and effect on process performance. It is important to understand the process equipment, conditions, and process steps that may be affected by critical utility performance, the variables that can adversely affect the critical utility performance, and the control measures in place to ensure critical utility performance and mitigate the risk of critical utility failure. The testing of these control measures throughout the expected process performance and parameter range should be demonstrated during Stage 2a.

Clean Rooms and Classified Areas

Controlled and/or classified facilities that support the parenteral manufacturing or testing process, including clean room and aseptic processing areas must be qualified. Controlled areas are those, without which the process would not reliably result in product meeting CQA specifications. Controlled and/or classified areas are designed to prevent microbiological contamination of the product or product contact surfaces. These areas are typically arranged in an increasing level of quality and protection from less controlled areas surrounding more critical areas. The core area where aseptic filling is performed is a Grade A or ISO 5 clean room spaces.

The requirements for classification of these areas may follow standards such as ISO 14644-1 and 2, as well as FDA Aseptic Processing Guidance (FDA, 2004) and EMA Annex 1 (EU Guidelines, 2008). Clean rooms and critical controlled areas should be designed to provide and maintain conditions required to manufacture sterile parenteral products, including proving environments with minimal or no levels of total particulates and viable microbiological contamination (ISO 14644-1, 2015).

Clean room and classified areas should be qualified under static and dynamic conditions. Static conditions set a baseline for clean room environment and performance without personnel and process step activity. As such this part of the qualification will demonstrate that the clean room and classified area is performing to design specifications. Qualification under dynamic conditions, with personnel and process step activity demonstrates that the clean room and classified areas can maintain environment and conditions required for process performance.

Clean room and classified area qualification should include high efficiency particulate air filters (HEPA) filter certification, velocity verification, differential pressure verification, airflow visualization (smoke studies), clean room wall, floor, and ceiling cleaning, and sanitization, temperature and humidity determination across room with full personnel activity and load, and automated control and alarm verification.

Airflow visualization or smoke studies are conducted to demonstrate that the air is flowing from areas of more cleanliness to areas of less cleanliness, first air is not disrupted during interventions and other activities and process steps, turbulence is not excessive, and airflow is unidirectional and in a downward direction (for vertical HEPA flow units). Air should flow in such a manner as to sweep particles away from product or product contact surfaces throughout the aseptic process, including interventions. Where interventions disrupt airflow, those interventions should be modified if possible. If they cannot be modified, then additional steps and actions, such as removal of open containers and exposed product contact surfaces, followed by decontamination of the affected area may be required. In that case, the decontamination process should be qualified, as well.

Airflow patterns in Grade A and ISO 5 areas generally should be shown to flow in a unidirectional manner from the source of the air, in most cases the HEPA filter face, to the exposed surfaces. Although a common term used is laminarity, it is difficult to demonstrate true laminar flow. Laminar flow is when a fluid flows in parallel layers without disruption. This rarely can be demonstrated in a typical clean room environment. Instead, what can and should demonstrated is a unidirectional airflow, where airflows in one direction, with no appreciable turbulence and back flow against that direction.

Airflow patterns in isolators and small environments may not exhibit unidirectional airflow due to the relatively small area and proximity of HEPA units to the surface, which may result in turbulence at hose surfaces. In these cases, a risk assessment of the area covered by the airflow can be used to determine if the airflow pattern is proper and adequate to meet process requirements.

Airflow velocity readings should be taken at or near the HEPA filter surface and closer to the working surface. US and European health authority guidance recommends velocities of 90 fpm (±20%) or 0.45 mps (±20%) at whatever point it is measured. However the airflow location and velocity may vary if it can be justified and can be shown to be sufficient to achieve unidirectional flows, remove particles form the area, and provide appropriate environmental conditions.

Clean room surface cleaning and decontamination should be qualified. Clean room surfaces, including walls, floors, panels, curtains, nonproduct contact machine surfaces, pass throughs, doors, piping, conduits, fixtures, and non-HEPA filter ceiling panels must be able to be cleaned and decontaminated. For purposes of this chapter decontamination, sometimes referred to as sanitization or disinfection, is defined as rendering the item or surface in capable of microbiologically contaminating sterile product or product contact surfaces.

The clean room cleaning and sanitization qualification should show that it can remove foreign materials, dirt, debris, and microbiological contamination from all surfaces including porous surfaces, spaces, gaps, or openings. It should have noted that clean room construction should be such that porous surfaces, spaces, gaps, and openings are minimized or eliminated. The qualification of the cleaning and decontamination usually includes a demonstration of the cleaning and decontamination disinfectant to destroy a panel of expected microorganisms on surfaces like the clean room application. It also should include challenge of the cleaning and decontamination process in situ in the clean room, using the methods and techniques employed during actual cleaning and decontamination—as performed by trained cleaning and decontamination personnel.

Component Preparation

The equipment and process used for the cleaning, preparation, sterilization, and where appropriate siliconization and depyrogenation of components used in the manufacturing of parenteral products, including vials, ampoules, syringes, stoppers, overseals/caps, syringe barrels, plungers, needle sheaths, etc. should be qualified and shown to be effective, reliable, and suitably designed. The qualification should show that the prepared components meet and maintain the quality specifications required for the manufacturing process. The wrapping, packaging, holding, hold times, and conditions should be established and qualified to ensure that components remain in a qualified state after preparation up to use in the manufacturing process. The qualification should show that particulates or extraneous materials are not added to the components during processing. The qualification should also show that the components are not damaged during preparation, handling, or transport. The packaging of components should be appropriate for clean room transfer into and use within the aseptic fill area. Conveyors, chutes, hoppers, bags, trays, etc. used for the transport of prepared components should be designed to allow for sterilization, decontamination, and aseptic handling to and within the aseptic filling area.

The sterilization and depyrogenaiton of parenteral product contact components should be qualified using appropriate microbiological challenge methods, including the use of biological indicators. Where components, such as glassware, are depyrogenated by dry heat, it is generally not necessary to qualify the sterilization of these components. This is due to the high heat levels used to depyrogentate, that exceed the levels needed to sterilize the components. Unless the components are heat sensitive, overkill methods are employed. In some cases, radiation and other means may be used to sterilize components. In these cases, the qualification should show that the composition, condition, or levels of leachables and extractables of the component is not adversely affected by the radiation. If chemical sterilization is used, for example, ethylene oxide, then the qualification should show that levels of residual chemicals and by products are within specification and will not adversely affect the quality for the parenteral product.

Care should be taken to show that stoppers and other plasticide components do not clump or occlude in a manner that would prevent the penetration of sterilizing media, such as moist heat from contacting all component surfaces. The sterilization process should also be qualified to show that air is removed during moist heat sterilization, where appropriate and that bags and wrapping materials are not damaged or rendered nonintegral during the sterilization process.

Product Sterilization and Filtration

Aseptically manufactured parenteral products can be sterilized by heat, filtration, and in some cases radiation or chemical means such as ethylene oxide.

Moist Heat Sterilization

Sterilization by heat, while the product remains in bulk, is usually performed when the product is heat stable, cannot be filtered, and where the product container cannot be terminally sterilized. Bulk sterilization is usually performed in a sealed tank or vessel, where the product temperature is raised to and held for a long enough time to achieve proper lethality. The time and temperature can be determined on an overkill basis, where the product is not temperature sensitive, or it can be based on bioburden of the product, where exposure to higher temperatures for extended time may be a product stability or efficacy concern.

Validation for the sterilization process involves qualification of the integrity of the holding vessel during and after sterilization, the integrity and effectiveness of venting filters, the temperature distribution within the holding vessel, the heating and cooling systems, the accuracy of the temperature monitoring and recording instrumentation, and the effectiveness of the sterile product transport system. Thermocouples placed throughout the bulk solution to demonstrate an even distribution of temperature and achievement of lethal conditions throughout the liquid bulk. Thermophilic biological indicators (Bis), selected specifically for resistance to moist heat, are used to demonstrate lethality throughout the vessel. *Geobacillus stearothermophilus* (bacillus *stearothermophilus*) is often used. Thermocouples and BIs should also be placed in the product and in the vessel dome, transfer lines, inlets, outlets, valves, traps, wells, and vent filters to demonstrate that the vessel is sterilized during the process. For a more complete discussion on moist heat sterilization, refer to PDA Technical Report No.1, Validation of Moist Heat Sterilization Processes (PDA, 2007).

The utilities that support the sterilization process should be qualified prior to the qualification for the sterilization process. These include but may not be limited to clean steam, cooling water, and compressed air.

Dry Heat Sterilization

Dry heat is used to sterilize aseptically manufactured parenteral products that cannot be filtered and cannot be exposed to moist heat. Examples include dry powders, oils, and ointments. In the case of dry powders, these may be a final product or part of a suspension. The steps for qualification of the dry heat process will be like that of moist heat. Bis should be selected for resistance to dry heat. *Bacillus subtilis* is often selected for its resistance to dry heat. For a more detailed discussion of dry heat sterilization, refer to PDA Technical Report No. 3, Validation of Dry Heat Processes used for Depyrogenation and Sterilization (PDA, 2013a).

Radiation and Ethylene Oxide

Nonheat methods, such as gamma radiation, e-beam radiation, and ethylene oxide are sometimes used to sterilize bulk powders. Where these methods are used, the sterilization process should be validated, and the equipment used for the process qualified. Care should be taken to ensure that radiation does not degrade the product and that the product remains sterile during handling and transfer (AAMI TIR33, 2005; ANSI/AAMI/ISO, 2013; ANSI/AAMI/ISO, 2017). Care should be taken to ensure that ethylene oxide residuals do not adversely affect or otherwise contaminate the products and that the product remains sterile during handling and transfer. Care should also be taken to ensure that powders do occlude or clump in a manner that does not allow for adequate exposure to the chemical sterilant (ANSI/AAMI ST41:2008/(R), 2012).

Sterile Filtration

Sterile filtration is probably the most widely used method for sterilization of aseptically processed liquid parenteral products. Filtration usually involves passing the product through a 0.22-micron porosity filter. In some cases, a second or redundant sterilizing filter is used. In some cases, a higher porosity prefilter is used to reduce bioburden or debris. The validation of the filtration process involves a bacterial retention challenge study. This study is usually performed by the filter manufacturer on actual product. The product is inoculated with large quantities of small diameter microorganism, usually *Brevundimonas diminuta* (*Pseudomonas diminuta*) is selected. Filters must demonstrate the capability to remove 10^7 μm mL^{-1} of product under anticipated production conditions for handling, sterilization, integrity test, flow, temperature, and usage. In addition, filter membrane integrity test parameters, conditions, and limits should be set for the product. For more information refer to PDA Technical Report No. 26 (PDA, 2008b) or ASTM F838-15-a (ASTM F838).

In addition, the sterilization of the filter and filter assembly must be validated and qualified. Moist heat sterilization may be accomplished by flowing clean steam through the filter and assembly while they are in place in the process transfer line for enough time, at a high enough temperature, and pressure to achieve lethal conditions. This is referred to as SIP (steam in place or sterilize in place). Where SIP is used, care should be taken to ensure that air is vented or removed from the assembly prior to or as part of the sterilization cycle. Care should also be taken to ensure that these conditions and parameters do not exceed manufacturer's recommendations or limits for the type of filter used. Care should also be taken to ensure that the differential pressure between the upstream and downstream sides of the filter do not exceed manufacturer's recommendations and limits. Care should also be taken to ensure that excessive condensate does not collect or accumulate in the filter, assembly, or product transfer lines during or after the sterilization cycle. The filter assembly should be loaded and positioned in the autoclave to allow for air venting and condensate drainage. Care should also be taken to ensure that the SIP-sterilized assembly and line remain in a sterile condition throughout the filtration process. The validation of the filter assembly sterilization involves the placement of thermocouples and biological indicators, like the method described in the previous section. For more information, refer to PDA Technical Report No. 61 (PDA, 2013b).

Moist heat can also be used to sterilize the filter and assembly in a sterilizer or autoclave. In this case, the filter is placed in the assemble and the open ends of the assemble are wrapped in a microbial barrier material. Care should be taken to ensure that the time, temperature, and pressure do not exceed manufacturer's recommendation and limits for the specific filter type. The filter assembly should be loaded and positioned in the autoclave to allow for air removal. Care should be taken to ensure that the filter assembly remans sterile throughout the removal from the autoclave, handling, storage, and connection to the aseptic processing line. For more information, refer to PDA Technical Reports No. 1 Validation of Moist Heat Sterilization and No. 48 Moist Heat Sterilization Systems (PDA, 2008c).

Disposable filters and assembly combinations may be provided to the user pre-sterilized. This is usually achieved through gamma radiation. In the case of autoclaved or gamma irradiated filters, the filter assembly is packaged or wrapped in a microbial barrier. The sterilized package or wrap should be qualified to show that it can maintain its microbial barriers during and after sterilization and storage. Time limits and conditions for storage of the sterilized package should be set and qualified.

Filling Equipment and Systems

The container/component loading, transport, filling, flushing, stoppering, and capping systems should be qualified to show that these systems perform as required, providing accurate fills and effective seals at production speeds, without adding microbiological, chemical, or particulate contamination. Each system and piece of equipment should be qualified as described in the equipment qualification section of the chapter. The qualification should include cleaning and sanitization of nonproduct contact surfaces, including sanitization that occurs after an intervention. Cleaning and sterilization of product contact parts and surfaces, and both static and dynamic airflow visualization studies as described in this chapter should be qualified. Systems that perform automated weight check, position verification, stopper placement verification, and container or unit rejection functions should be qualified. Nitrogen flushing systems and functions should be qualified and tests to ensure the quality and purity of the nitrogen flush should be confirmed.

The cleaning and sterilization of filler product contact parts, including hosing, fill nozzles/needles, pumps, lines, valves, connectors, stoppering and plunger bowls, tracks, and placement assemblies should be qualified and validated. This would include the qualification of parts wrapping, storage, and cleaning process steps and time limits.

The decontamination and transfer of sterilized parts, materials, items, utensils, carts, and components into the clean room area and then into the Grade A critical fill area should be qualified and validated. This includes disinfectant qualification studies in the laboratory that show the effectiveness of the disinfectant to destroy anticipated microbiological contamination, as well as in situ testing on packaging, wrapping, and transfer item surfaces.

The transfer of materials into the Grade A environment and the setup of sterile filler parts and equipment will further be tested during the aseptic process simulations, as inherent or routine interventions. However the aseptic process simulations should not be the only method for qualifying these steps. Instead media fills should be part of an overall holistic aseptic process validation, as described in the aseptic process simulation chapter of this book.

Inspection, Labeling, and Secondary Packaging Systems

The equipment and systems used for packaging operations, including inspection, labeling, cartooning, insert placement, coding, and case packing systems should be qualified at anticipated production line speeds. Where these processes are automated, the control and monitoring systems should be qualified. The inspection of filled units for defects, including visible particulates, improper or inadequate seals, bad cap crimps, cracks, and other evidence of nonintegral containers, blemishes, foreign materials, and incorrect components should be qualified to show that representative defects are identified and removed, and an excessive number of acceptable units are not rejected. Where manual or semimanual visual inspection occurs, lighting and conveyor speeds are of importance. Automated visual inspections systems should be shown to remove defects to at least the same extent and success as manual systems.

Label coding, label application, and label reading systems should be qualified under production speeds and conditions. Secondary packaging systems, including transport, accumulators, product insertion, insert insertion, cartoning, secondary labeling, coding, sealing, and case packing, as well as associated monitoring and control systems should be qualified.

Lyophilization

The equipment, systems, and utilities that perform and support the lyophilization for parenteral products should be qualified and validated. These systems include the lyophilizer, transport, and loading equipment/systems. The cleaning and sterilization of the lyophilizer interior should be qualified. The integrity of the lypohilizer, including venting filters, should be tested during the qualification. The capability of the lyophilizer shelves to seat stoppers and seal vials should be tested. The lyophilization cycle should be qualified. The nitrogen distribution system and the quality of the nitrogen used to break the vacuum should be qualified. The operation of condensers and compressors should be tested. The control, monitoring, recording, and alarm systems should be tested and qualified. The actual lyophilization of product should be confirmed during the PPQ for that product. Those parts of the lyophilization operation and cycle where parenteral product is exposed should be tested during aseptic process simulations.

Material Storage, Handling, and Transport

The handling and storage of finished parenteral products that may have an impact on the quality of the product, including those systems that if improperly designed or operated may damage product, maintain product temperature and pressure during storage and transport, track product status, segregate product, maintain records of product status, and distribution and secure product should be qualified. Where these systems are automated, the control, monitoring, and recording instrumentation should be qualified. Temperature sensitive precut maintenance, including cold chain and shipping should be validated. Cold rooms, refrigerators, freezers should be qualified. The containers used for shipping should be qualified to maintain appropriate environmental conditions during transport, shipping, and storage. Active shipping and transport containers, where mechanical systems are sued to maintain environmental conditions should be qualified. Security systems should be qualified.

Testing Laboratories

The facilities, areas, instruments, systems used for testing of product, materials, and components, including in-process testing, monitoring, finished product testing, validation testing, and stability testing should be qualified. The analytical and microbiological test methods should be validated or otherwise shown to be reliable. These systems include fume hoods, biosafety cabinets, laminar flow hoods, sterility test isolators, autoclaves, sterility, refrigerators, incubators, stability chambers, testing instruments, and automated data collection and recording systems.

Additional Parenteral Processes

The following are important parenteral manufacturing processes, that due to the scope and limitations of this chapter cannot be covered in detail. Any one of the topics commands much discussion and can be further explored in the noted references.

Cleaning Validation

The process for cleaning product contact surfaces should be qualified and validated to show that dirt, foreign materials, and residual product have been removed or reduced to a noncritical level. The acceptance criteria for product residual should be based on clinical effect of residual on patients. The analysis should include those products that are most difficult to remove, those surfaces most difficult to clean, and those residuals most difficult to detect. For more complete discussion refer to PDA Technical Report No. 29 (PDA, 2012a).

Computer System Validation

The computerized and programmable systems used to automate, control, monitor, alarm, and secure process equipment, facilities, utilities, systems, and instruments should be qualified and validated. In some cases, the validation will include a review of source code. In some cases, the validation will include functional testing and a review of security measures. A risk assessment can be used to determine what level of testing and qualification is needed. For more complete discussion ref to ISPE GAMP (ISPE).

Terminal Sterilization

Terminal Sterilization is usually the exposure of sealed parental finished product to lethal levels of moist heat in an autoclave or sterilizer. The validation of terminal sterilization includes the qualification of the sterilizer and clean steam system, cycle parameters, loading patterns, monitoring devices, and product closure seals.

Container Closure Integrity

The integrity of the sealed container, vial, syringe, or ampoule should be qualified and the inspection and testing methods for container closure integrity should be qualified. In some cases, container closure integrity should be confirmed through inspection of all finished product units. In some cases, qualification of the container closure design and function linked to statistical in process or finished product testing is sufficient. A risk assessment can be used to determine what level of testing and qualification is needed.

Periodic Assessment and Requalification

As noted in other chapters, Stage 3 of the process validation life cycle is continued or ongoing process verification. This stage continues for the life of the process. It is designed to identify process variables that may not have been addressed during process design, because they were missed or appeared after Stage 1 and 2. Stage 3 provides an opportunity to improve the process by now addressing those variables.

Stage 3 confirms, through monitoring, analysis, and evaluation, whether the process remains in a qualified state. It is also important to confirm that the equipment, facilities, and systems qualified in Stage 2a remain in a qualified state. Periodic or event-driven requalification may not be enough if the qualification does not consider variables that were not addressed during the initial qualification. Instead it may be more beneficial to periodically access the systems and based on the outcome of that assessment determine what if any actions are needed, including but not limited to requalification.

The schedule for periodic assessment should be established using a risk-based approach, taking into consideration the impact of system failure on crucial product attributes, as well as system complexity, prior knowledge, and system history. Those systems with direct impact on product sterility, such as sterilizers may require more frequent assessment and annual requalification as a default schedule.

The periodic assessment should also be risk-based, taking into consideration the performance of the system, failure investigations, monitoring results and data, adherence to preventive maintenance schedules, preventive maintenance indicators, open work orders, corrective actions and preventive actions, change control, nonconformity reporting, and changes in regulatory submission and a health authority expectation. The result of the assessment may indicate that no actions are needed, some testing or procedural changes or controls are required, or requalification or new qualification testing is required.

Stage 2b—PPQ or Process Performance Qualification

PPQ sometimes referred to as Stage 2b is that part of PQ that involves confirmation batches. It includes PPQ confirmation batches. PPQ demonstrates that the process using the qualified equipment, facility, and utilities can produce product meeting the desired and required quality specifications. PPQ is most associated with what one might refer to as traditional process validation, the running of three successful process batches. However there are some important difference. The acceptance criteria for PPQ involves a careful analysis of data from samples obtained form process batches, rather than merely the successful running of these batches. The number and configuration of samples must be scientifically justified and statistically sound.

Number of Batches

As discussed earlier, throughout the FDA 1987 process validation guidance, there are references to finished product testing to confirm the performance and capability of the process. This led to a reliance on the running and evaluation replicate batches as a primary means to validate the process. However, this has proven to be less than effective, as the product specifications and test criteria are set to confirm product acceptability and not process capability. The realization that running of successful replicate batches does not validate the process, represents the most significant difference in the 1987 process validation approach and the 2011 FDA guidance approach. This being that it is not the successful running of batches that defines the criteria for acceptance, but the analysis of the data obtained from running those batches.

The number of batches and samples within each batch must be justified and have a sound statistical and scientific basis. The justification can also be based on other criteria, such as prior knowledge and scientific principle. Samples from multiple batches should be required to demonstrate interbatch consistency. EU Annex 15 suggests a default minimum number of three batches. However there must be sound scientific and statistical basis for the number of samples and batches even if three is the number selected. Consideration should be taken for both inter- and intrabatch consistency (EU Guidelines, 2015).

The number and selection of samples for that analysis is a matter of determining the confidence required to affirm process performance and the coverage needed to obtain that confidence. Confidence is based on the severity or impact of a process failure on one or more product CQAs.

CQAs define the quality aspects of the product. A CQA is a feature of the product, without which it is no longer meets user needs. CQAs are often defined in the Quality Target Product Profile (QTPP) (ICH Q8) of a product. The QTPP describes the design criteria for the product and should therefore form the basis for development of the CQAs, CPPs, and control strategy (FDA, 2015). CQAs are defined in ICH Q8 as "a physical, chemical, biological, or microbiological property or characteristic that should be within an appropriate limit, range, or distribution to ensure the desired product quality" (ICH, 2009).

The more the impact on a CQA, the more confidence may be needed, and therefore the number of samples and batches for PPQ. However statistical coverage is only one of many tools available for establishing process confidence. For instance, there are also scientific principles, experimentation, and prior knowledge, which can support or help determine confidence and coverage levels.

Coverage levels or number of batches and samples will also depend on the extent, robustness, and outcome of Stage 1 process design studies, including engineering studies or similar processes. Likewise the results of the PPQ sampling evaluation will help determine the number of batches, samples, and evaluation needed during the enhanced sampling phase of Stage 3—continued or ongoing process verification.

One practical method of deriving the number of batches to be include into PPQ study we could offer is utilization of statistical Power. We will briefly discuss this concept so that Process Validation practitioners may use it in design of their protocols. Power is an important property of any hypothesis test because:

- It indicates the likelihood that you will find a significant effect or difference when one truly exists.
- Power is the probability that you will reject the null hypothesis in favor of the alternative hypothesis.

Statistical power is defined as the probability of not missing an effect, due to sampling error, when there really is an effect there to be

Table 8.2 Cohen's Rules of Thumb for Effect Size

Effect Size	Correlation Coefficient	Difference Between Means
Small effect	$r = 0.1$	$d = 0.2$ standard deviations
Medium effect	$r = 0.3$	$d = 0.5$ standard deviations
Large effect	$r = 0.5$	$d = 0.8$ standard deviations

found. Power is the probability (prob $= 1 - \beta$) of correctly rejecting H_o when it really is false. Calculation of statistical Power depends on the following factors:

- the sample size;
- the level of statistical significance required; and
- the minimum size of effect that it is reasonable to expect.

The sample size and statistical significance are easily determined based on known process variability. To determine minimum size effect typically Cohen "Rule of Thumb" for effect of size is used (Cohen, 1988, 1992) (Table 8.2).

The following conventions and decisions about statistical Power are considered:

- Acceptable risk of a Type II error is often set at 1 in 5, that is, a probability of 0.2.
- The conventionally uncontroversial value for "adequate" statistical power is therefore Set at $1 - 0.2 = 0.8$.
- It is often regarded the minimum acceptable statistical power for a proposed study as being an 80% chance of an effect that really exists showing up as a significant finding.

Then ANOVA is used to determine the number of batches based on known process variability that is derived from CQA results of clinical batches, engineering batches, design of experiments batches, and scale-up batches. An example ANOVA for a Number of Batches $N = 3$ (refer to Fig. 8.2) shows that even with 10 samples we have a high power to detect close to 2 units difference from the mean. In addition the curve shows us that with current standard deviation we can determine up to 3 units with very high power. Therefore, it means 10 samples has a statistical power to detect differences from what we know about studied population. Not more and not less.

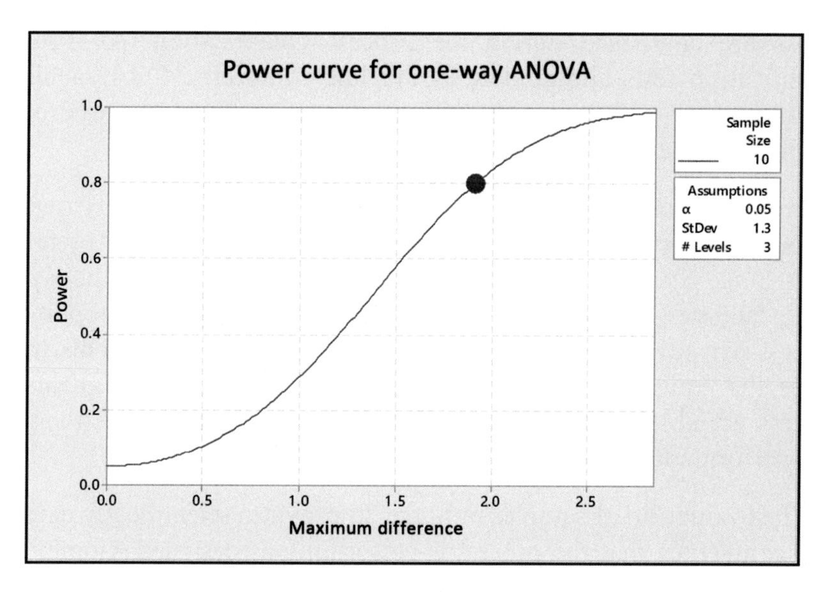

Figure 8.2 An example ANOVA for a number of batches N = 3.

Based on this example one can determine the number of batches that is based on a number of samples needed to detect currently known process variability.

Conclusion

Stage 2 of the Process Validation Life Cycle is referred to as the Process Qualification Stage. It is the qualification through testing of the equipment, facilities, systems, instruments, and processes that support the overall parenteral product manufacturing process. Process Qualification should build on information obtained from Stage 1, Process Design. It is Process Design that determines the process needs, variables, and control strategies to address those variables and ensure process capability. Process Qualification then tests the assumptions by which the process was designed to provide further confidence of prolonged, reliable process performance.

Using a Line of Sight approach, Process Qualification tests, data acquisition, and analysis are linked to the establishment and maintenance of critical product quality attributes (CQAs), these CQAs are the objective of the overall manufacturing process. The CPPs and conditions under which the process will perform properly and result in

CQAs are confirmed during the Process Qualification. All Process Qualification tests and studies should link to specific CQAs achievement, and all CQAs should be covered by one or more Process Qualification tests or studies.

Process Qualification can be divided into equipment/facility qualification, sometimes referred to as Stage 2a and PPQ, sometimes referred to as Stage 2b. The objective of Stage 2a is to show that the equipment, facilities, utilities, systems, and instruments used to support the process can do so and will perform in a reliable manner. This minimizes the equipment, facilities, utilities, systems, and instruments as process variables and allows for more effective evaluation of the process performance in Stage 2b.

The evaluation of samples in Stage 2b provides scientifically defendable confidence that the process is performing as designed. Combining that knowledge with the well thought out and tested design from Stage 1 and the reliability of equipment and systems established in Stage 2a, should provide confidence that the process will result in product meeting quality specifications and standard and can be committed to commercial manufacturing. The testing and sampling regiment established during Process Qualification can then be used as the basis for Stage 3, continued Process verification sampling, testing, and evaluation.

References

AAMI TIR33, 2005. Technical Information Report: Sterilization of Health Care Products—Radiation—Substantiation of a Selected Sterilization Dose—Method VDm.

ANSI/AAMI/ISO 11137-2, 2013. Sterilization of Health Care Products—Radiation—Part 2: Establishing the Sterilization Dose.

ANSI/AAMI/ISO 11137-3, 2017. Sterilization of Health Care Products—Radiation—Part 3: Guidance on Dosimetric Aspects of Development, Validation and Routine Control.

ANSI/AAMI ST41:2008/(R), 2012. Ethylene Oxide Sterilization in Health Care Facilities: Safety and Effectiveness.

ASTM F838 - 15a Standard Test Method for Determining Bacterial Retention of Membrane Filters Utilized for Liquid Filtration.

Cohen, J., 1988. Statistical Power Analysis for the Behavioral Sciences. Routledge, ISBN: 978-1-134-74270-7.

Cohen, J., 1992. A power primer. Psychol. Bull. 112 (1), 155–159. Available from: https://doi.org/10.1037/0033-2909.112.1.155. Available from: 19565683.

EU Guidelines, 2008. EU Guidelines to Good Manufacturing Practice Medicinal Products for Human and Veterinary Use Annex 1, Manufacture of Sterile Medicinal Products (corrected version) EudraLex The Rules Governing Medicinal Products in the European Union, Volume 4, Brussels, 25 November 2008 (rev.)

EU Guidelines, 2015. EU Guidelines for Good Manufacturing Practice for Medicinal Products for Human and Veterinary Use Annex 15: Qualification and Validation, EudraLex Volume 4, Brussels, 30 March 2015.

FDA, 1987. Guidelines on General Principles of Process Validation, Food and Drug Administration.

FDA, 2004. Guidance for Industry, Sterile Drug Products Produced by Aseptic Processing—Current Good Manufacturing Practice, U.S. Department of Health and Human Services Food and Drug Administration, Center for Drug Evaluation and Research (CDER), Center for Biologics Evaluation and Research (CBER), Office of Regulatory Affairs (ORA), September 2004, Pharmaceutical CGMPs.

FDA, 2011. Guidance for Industry Process Validation Guidance: Principals and Practices, US Department of Health and Human Services, Food and Drug Administration.

FDA, 2015. Presentation - How to Identify Critical Quality Attributes and Critical Process Parameters Jennifer Maguire, Ph.D. Daniel Peng, Ph.D. Office of Process and Facility (OPF) OPQ/CDER/FDA 1 FDA/PQRI 2nd Conference North Bethesda, Maryland, 6 October 2015.

ICH, 2009. Pharmaceutical Development, Q8(R2), Step 4 version, August 2009, International Conference for Harmonization.

ISO 14644-1, 2015. Cleanrooms and Associated Controlled Environments—Part 1: Classification of Air Cleanliness by Particle Concentration Standard International Organization for Standardization, 15 December 2015.

ISPE, A Risk Based Approach to Compliant CxP Computerized Systems GAMP Current Version.

ISPE, 2019. Baseline Guide: Commissioning and Qualification Revision 2, International Society for Pharmaceutical Engineering.

Long, M., 2013. Risk and Statistics Serve as Tools for Solving Variation Riddles, PDA Letter, April 2013.

Maki, D., Rhame, F., Mackel, D., Bennett, J., 1976. National epidemic of septicemia caused by contamination of intravenous product. Am. J. Med. 60.

PDA, 2007. Technical Report No. 1 Validation of Sterilization Cycles, PDA, Bethesda.

PDA, 2008a. Comments for Federal Registry Part 211 Changes.

PDA, 2008b. Technical Report No. 26, Revised 2008, Sterilizing Filtration, of Liquids, PDA, Bethesda.

PDA, 2008c. Technical Report No. 48 Moist Heat Sterilizer Systems: Design, Commissioning, Operation, Qualification and Maintenance, PDA Bethesda.

PDA, 2012a. Technical Report No. 29 Points to Consider for Cleaning Validation, PDA, Bethesda.

PDA, 2013a. Technical Report No. 3 (Revised 2013) Validation of Dry Heat Processes Used for Depyrogenation and Sterilization, PDA Bethesda.

PDA, 2013b. Technical Report No. 61 Steam in Place, PDA, Bethesda.

Further Reading

PDA, 2012b. Technical Report 60, Process Validation—Risk Based Lifecycle Approach.

CHAPTER 9

Process Validation Stage 3: Continued Process Verification

Igor Gorsky
Senior Consultant, ConcordiaValsource LLC., Downingtown, PA, United States

Introduction

This chapter will discuss Stage 3 of Process Validation Program—Continued Process Verification (CPV). First of all, it is important to understand regulation requirements that are behind the concept of continued verification of Critical Process Parameters (CPP) and Critical Quality Attributes (CQA). Before we start this discussion, one should distinguish between "continued" and "continuous" designation in the definition of this stage of Process Validation. "Continued" refers to ongoing, periodic review that provide "continual assurance that the process remains in a state of control (the validated state) during commercial manufacture (FDA, 2011)," while "continuous" describes an alternative approach to process validation and manufacturing in which manufacturing process performance is continuously monitored and evaluated (ICH, 2009) typically using process analytical technology methods (FDA, 2004). The reason we need to make this distinction is because unfortunately we continually observe mislabeling of these terms in technical literature and feel that clear definition would be useful to a reader.

The basis for CPV as it is discussed in FDA's Process Validation Guidance for Industry lies in the Code of Federal Regulations Section §211.180(e) under subparts J—Records and Reports. This section in subsections (e) and (f) states the following:

"(e) Written records required by this part shall be maintained so that data therein can be used for evaluating, at least annually, the quality standards of each drug product to determine the need for changes in drug product specifications or manufacturing or control

Principles of Parenteral Solution Validation. DOI: https://doi.org/10.1016/B978-0-12-809412-9.00008-3

procedures. Written procedures shall be established and followed for such evaluations and shall include provisions for:

1. A review of a representative number of batches, whether approved or rejected, and, where applicable, records associated with the batch.
2. A review of complaints, recalls, returned or salvaged drug products, and investigations conducted under 211.192 for each drug product.

(f) Procedures shall be established to assure that the responsible officials of the firm, if they are not personally involved in or immediately aware of such actions, are notified in writing of any investigations conducted under 211.198, 211.204, or 211.208 of these regulations, any recalls, reports of inspectional observations issued by the Food and Drug Administration, or any regulatory actions relating to good manufacturing practices brought by the Food and Drug Administration (Code of Federal)." The CPV program recommended in FDA Guidance based specifically on this section of Code of Federal Regulation (CFR). In its guidance FDA goes further to state that "data collected" on product and process "should include relevant process trends and quality of incoming materials or components, in-process material, and finished products." FDA continues stating that "data should be statistically trended and reviewed by trained personnel" and "the information collected should verify that the quality attributes are being appropriately controlled throughout the process (FDA, 2011)."

As written in FDA's Guidance to Industry, "Process Validation: General: Principles and Practices (January 2011)," process validation is defined as the collection and evaluation of data, from the process design stage through commercial production, which establishes scientific evidence that a process is capable of consistently delivering quality product. Process validation involves a series of activities taking place over the life cycle of the product and process. The FDA Guidance to Industry further describes process validation activities in three stages. The first two stages were discussed in the previous chapters while this chapter concentrates on Stage 3—CPV that should provide an ongoing assurance which is gained during routine production that the process remains in a state of control. Furthermore FDA affirms "a successful validation program depends upon information and knowledge from product and process development. This knowledge and understanding is the basis for establishing an approach to control the manufacturing process that results in products with the desired quality attributes (FDA, 2011)." FDA expects parenteral product manufacturers are able to answer a few pivotal questions prior to commercial distribution of products:

- Does manufacturer understand the sources of variation in their processes?
- Does manufacturer have ability and tools to detect the presence and degree of that variation?
- Does manufacturer understand the impact of variation on the process and ultimately a degree of an impact on product attributes that may affect patient?
- Can manufacturer control the variation in a manner proportionate with the risk it represents to the process and ultimately the product?

The 21st century compliance, with its three-way emphasis on process design, process qualification, and CPV, treats validation as an integrated and risk-based life cycle activity, whose aim is to identify and minimize sources of variability within the manufacturing process, and to quantify and manage residual risk in a proportionate manner. This represents an improvement over the traditional approach to compliance, whereby validation was viewed as a standalone activity, with little account being taken of formalized process monitoring during the operational phase. This chapter will address CPV aspects, recommending one element of the parenteral products manufacturers' response to emerging regulation and manufacturer's responsibility for a Process Validation program. The primary objectives of CPV are:

- to demonstrate that manufacturing processes remain in an ongoing state of control,
- to facilitate the acquisition and communication of process understanding, and
- to identify specific opportunities and targets for process improvement.

A program of CPV provides a means to ensure that processes remain in a state of control following the successful Process Performance Qualification (PPQ) stage and during the commercial manufacture stage. The information and data collected during Stages 1 (Process Design) and 2 (Performance Process Verification) sets the stage for an effective control strategy in routine manufacturing and a meaningful CPV program. The understanding of functional relationships between process inputs and corresponding outputs established in earlier stages is fundamental to the success of the CPV program. Evaluating the performance of the process identifies problems and determines whether action must be taken to correct, anticipate, and prevent problems so that the process remains in control.

Continued monitoring of process variables enables adjustments to inputs covered in the scope of a CPV plan. It compensates for process variability, ensuring that outputs remain consistent and meet expectations and specifications. Since all sources of potential variability may not be anticipated and defined in Stages 1 and 2, unanticipated events or trends identified from continued process monitoring may indicate process control issues and/or highlight opportunities for process improvement. Science and risk-based tools help achieve high levels of process understanding during the development phase, and subsequent knowledge management across the product life stages, facilitates implementing continuous monitoring. There are two components to a CPV program:

1. Initial or short-term CPV (designated in technical literature as Stage 3a)
    a. This is an initial stage that occurs at the commencement of commercial manufacture. Typically, during this stage, data are being collected and used to calculate control (alert) limits after completion of a statistically appropriate number batches from which a statistically appropriate and risk-based number samples have been taken.
    b. These limits are used as the starting points for the long-term monitoring CPV Plans known as Stage 3b.
    c. The relative risk of the Critical Material Attributes, CQA, or In-Process Control Parameters (IPCs) being monitored determines the number of samples taken.
    d. For new products and processes, Stage 3a is prospective in nature as there generally may not be enough data to move from Stage 2 (PPQ) into commercial manufacture (Stage 3) to calculate statistically relevant limits or targets.
    e. For Legacy Products, Stage 3 a is generally retrospective, utilizing data for batches that have already been produced to determine the limits to be applied in Stage 3b.

2. Long-term CPV (designated as Stage 3b)
   a. This is the long-term monitoring stage of commercial manufacture that continues through a life cycle of the product or until significant changes are made that will require reevaluation of the Process Control Strategy and possible reexecution of Stage 2 (PPQ). During this stage of CPV, limits and targets generated during Stage 3a are used to monitor products and processes.
   b. Excursions from those limits and targets trigger assessments, actions, and plans for continuous improvement. The review period is risk-based and limits are evaluated and adjusted if required as a part of the review process.

This chapter may serve as a guidance recommendation for defining Stage 3a and 3b plans in support of CPV activities and to complement the existing Pharmaceutical Quality Management for parenteral products.

Definitions Typically Used in Process Validation

We shall list some of the typically used definitions for the sake of a future discussion. These definitions are listed in ICH,[1] FDA,[2] EMA,[3] MHRA,[4] HPRA,[5] CFDA,[6] PMDA,[7] Health Canada, WHO, and other regulatory agencies, as well as international standards organizations (ISO and ASTM) and industry professional organizations (PDA and ISPE). Some of the wording of these definitions may be slightly different as they are listed in difference guiding documents; however, main concepts are the same and are acceptable to be used. It is beneficial to list them prior to start of this chapter to establish baseline of our discussion.

- Acceptance Criteria is a set of conditions/specifications that a product must meet to be accepted by product's release specifications.
- Capability of a process is an ability of a process to produce a product that will fulfill the requirements of that product which can be defined in statistical terms. (ISO 9000:2005).

[1] International Congress for Harmonization.
[2] US Food and Drug Administration.
[3] European Medicinal Agency.
[4] Medicines and Healthcare products Regulatory Agency.
[5] Health Products Regulatory Authority.
[6] Chinese Food and Drug Administration.
[7] Japanese Pharmaceuticals and Medical Devices Agency.

- Commercial manufacturing process is a process resulting in commercial product (i.e., drug that is marketed, distributed, and sold or intended to be sold).
- Concurrent Data Set is a set of data that has been produced from an approved assay or process during the most recent specified time period.
- Concurrent release is release for distribution of a batch of a finished product that was manufactured following a qualification protocol, that meets the batch release criteria established in the protocol, but before the entire qualification study protocol has been executed.
- Confidence Level is a level that is the long-run proportion of intervals constructed in this manner that will include at least the specified proportion of the sampled population.
- CPV is a program of an ongoing assurance, gained during routine production that the process remains in a state of control.
- Continuous data are data that are contained or measured on a continuous, infinite scale which can take any value and can be subdivided in a meaningful manner and is typically obtained from scales, gauges, sensors, etc.
- Control Limits are limits used for trending which are normally distributed data, where the upper control limit is equal to center line (CL) or mean plus 3 standard deviations (SD) and the lower control limit is equal to CL minus 3SD.[8]
- Coverage Probability is a proportion of items in a population lying within a statistical tolerance interval.
- Discrete Data are a set of data values with unconnected data points, often a count or score.
- Historical Data Set is a selection of data that are regarded as standard and under control for a particular process or test. This data set is used to calculate the control limits in trending for CPV.
- Nonparametric Data are data that do not meet the assumptions of normality and, consequently, have no assumption of an underlying normal distribution
- Normally Distributed Data are a theoretical frequency distribution for a set of variable data, usually represented by a bell-shaped curve symmetrical about the mean, also named *Gaussian distribution*.

[8] For nonnormally distributed data, control limits may be established based on a frequency distribution of the historical data set, for example, utilizing percentiles.

- Out of Specification (OOS) is any test result that falls outside predefined specifications or acceptance criteria established in the drug's filed application (in United States typically NDA, ANDA, or BLA), or official compendia, or limits for release set by the manufacturer, that is tested through all stages of testing allowed by the specifications.
- Out of Trend is any test result within a predefined specification or acceptance criteria established in the drug's filed application (in United States typically NDA, ANDA, or BLA), or official compendia, or limits for release set by the manufacturer, which may be atypical of the testing history, or is indicative of the potential for a batch to fail meeting the standards of quality established in the specifications during its shelf life.
- Outlier is an unusually large or small observation that may have a disproportionate influence on statistical results, such as the mean, which can result in misleading interpretations.
- Parametric data are data that meet the assumptions of normality and, consequently, have there is an assumption of an underlying normal distribution.
- % Relative standard deviation (%RSD) or % coefficient of variation (%CV) of a series of measurements is statistic that is measure by using the following equation: $\%RSD = \frac{(Standard\ Deviation)}{Mean} \times 100\%$.
- Process Design is a definition of the commercial manufacturing process based on knowledge gained through development and scale-up activities.
- Performance indicators are measurable values used to quantify quality objectives to reflect the performance of an organization, process or system, also known as *performance metrics* in some regions (ICH Q10).
- Parts per million (PPM) is a statistics typically used in Process Capability and Performance calculations.
- Process Performance Index (Ppk) is a capability index of process performance that is adjusted for noncentered data. The ability of a process to produce product that meet specifications.
- PPQ is Stage 2 of Process Validation which combines the facility, utilities, equipment qualification, and the trained personnel with the commercial manufacturing process, control procedures, and components to produce commercial batches.
- Process qualification is confirmation that the manufacturing process as designed is capable of reproducible commercial manufacturing.
- Process validation is the collection and evaluation of data, from the process design stage through commercial production, which

establishes scientific evidence that a process is capable of consistently delivering quality products.

- Proportion is a part, share, or number considered in comparative relation to a whole.
- Quality is the degree to which a set of inherent properties of a product, system, or process fulfills requirements (ICH Q9).
- State of control is a condition in which the set of controls consistently provides assurance of continued process performance and product quality (ICH Q10).
- Statistical Tolerance Intervals[9] are intervals determined from a random sample in such a way that one may have a specified level of confidence that the interval covers at least a specified proportion of the sampled population.
- Statistical Tolerance Limit is a statistic representing an end point of a statistical tolerance interval.
- Statistical Tolerance Limit Factor (*k*) is a positive value that accounts for confidence level of a tolerance interval, and sampling errors in sample average and standard deviation for a specified confidence level and coverage probability.
- Variance is a numerical value used to indicate how widely individuals in a group vary.
    - Common-cause variance is inherent or random variability in the process and or analytical method which may only be reduced by changing the process or method of analysis itself.
    - Special-cause variance is resulting from factors that are not inherent to the process or method and which are not always present.[10]

Determining When Continued Process Verification Starts

Let us talk about a title question of this section. How do we determine when CPV starts? Well, isn't it obvious? After PPQ is completed, of course. How do we decide on how many batches we will include into early stages of CPV? Do we have enough knowledge about process variation to establish a practical program? These are valid questions

[9] Statistical tolerance intervals can be either one-sided, in which case they have either an upper or a lower statistical tolerance limit, or two-sided, in which case they have both.

[10] A process or method of analysis is considered in control if there are no special causes affecting it. Once special causes are removed or corrected the process or method of analysis should be considered in control.

and these questions can only effectively be answered when proactive planning takes place.

It is interesting to note that the question of introduction of CPV program is very similar to introduction of Quality Management System (QMS) into Research and Development organization. Typical R&D organization although aspiring to, one day, get a commercial product approved will historically skeptically look at introduction of QMS during manufacturing of clinical batches. The author once asked two regulators from different regulatory agencies from two different parts of the world—when QMS should be instituted. One said the following—"in regard to the timing at which a QMS should be implemented I think the question is a bit misleading in that a QMS should be in place from the outset as it is effectively serves as the framework within which Quality Assurance (QA) (and other functions) operate. However, a QMS should not be a fixed entity that has the same stringency throughout the life cycle phases, rather it becomes more stringent as the product advances forward. Therefore one should not commence making a product for human (clinical) use without having a QMS in place but that QMS will change over time. Consider the alternative: one manufacture clinical lots then applies for Regulatory approval to administer that product to humans (clinical studies) but have no documentation of independent review of the product. The regulatory agency would not look favorably on such firm before rejecting the application." The other one said—"it is generally understood that the QMS 'shall' be implemented when the commercial operations begin. That said, a company probably 'should' have the QMS or similar mechanism in place to manage the data and operations in Phase III. Before that, there should still be a mechanism to manage the data and manufacturing operations, but it is much softer and less clearly defined. This would be the minimal standard in my opinion."

This was an interesting comparison of two regulators' opinions as one would like to see a significant quality system from the beginning of the product/process development, while the other states that firms may have quality systems at Phase 3 Clinical studies, for sure.

We side with the first regulator. The FDA's Process Validation Guidance was revised precisely for reason of establishing, documenting, developing, analyzing, and evaluating Process Validation programs at the relevant stages of development, at least at the point of finalization of the process. This is where one should start thinking about CPV. The finalization of the process may occur at any phase of clinical study program; therefore one should be ready to implement CPV as early in the process development as possible. It is absolutely certain that by the time PPQ is completed and launch batches must me manufactured, those launch batches should already be a subject of CPV.

Therefore typically, CPV program should be considered just prior to implementation of Stage 2b—PPQ. As CPV consists of Stage 3a, followed by Stage 3b it is important to devise a structured program consisting of policy, standard, Standard Operating Procedure, CPV Plan, and Periodic CPV Reports. Introduction of a structured system is practical and proactive approach for CPV program. It is recommended to employ a multidisciplinary team when instituting such a program with an early involvement of the QA. The Stage 3a begins with an approved CPV Plan. Stage 3a ends when the sufficient number of batches have been produced and analyzed to determine process inherent variability. Typically Critical Quality Attributes shall demonstrate a state of a statistical control and may suffice to determine process variability and to proceed to Stage 3b. Stage 3a is typically applied to products that are just completed PPQ studies. Preparation for Stage 3a plan should be established prior to PPQ studies, provided significant process knowledge exists from Research and Development and Engineering batches. Products may reenter Stage 3a (after being in Stage 3b) if necessary.

Another set of prerequisites without which CPV should not start as listed below in Table 9.1. A system of obtaining these data must be installed at the firm so that data analyses could start immediately upon data attainment. It should be noted that if the data are entered manually, it should be independently checked, and independent verification must be documented. There are number of possibilities exist that guard data integrity. Some of these are listed as follows:

- manual data entry with independent check and hard copy documentation;
- qualified third-party enterprise spreadsheet with Part 11 and Annex 11[11] compliance; and

[11] FDA—21CFR Part 11.
EMA—Eudralex Volume 4, Annex 11.

Table 9.1 Example CPV Prerequisites	
Data Sources/Documents	**Data Description and Format**
Quality Target Product Profile	Critical Quality Attributes and Rationale
Process Flow Diagram, Batch Record, etc.	Process Parameters/Material Attributes
Certificate of Analysis (C of A)	Quality Attributes
Process Performance Qualification Report	All tables in Worksheet or Statistical Software format tabulated into columns (e.g., each batch, run, test) in an individual column.
Summary of All Available Results from Every Batch Manufactured C of A	All tables in Worksheet or Statistical Software format tabulated into columns
Stability Studies	Time zero data for all specifications for all batches on room temperature stability (data in Worksheet or Statistical Software column format—column for each batch, each attribute/specification)

- qualified Part 11 and Annex 11 compliant Data Acquisition and Analyses software package;

Legacy Systems Versus New Systems

Presented in this chapter, is a multistep process general and approach for CPV which also include some definitions for a statistically sound Stage 3a and Stage 3b CPV plans (Fig. 9.1).

There are many factors to consider in the creation of CPV plans, including:

- Is the plan for a new or legacy product or process?
- Is there an understanding of material attributes and quality attributes?
- Is there an understanding of the relationship between process parameters and quality attributes (i.e., Does the product or process have an approved criticality analysis report?)
- Are the data continuous and normally distributed or is it discrete?
- What statistical tools should be used in the plan?
- Is the plan for a low volume product?
- What control charting rules should be employed?
- How often should control limits (once established after Stage 3a) be reviewed?

First when creating a CPV plan one must review existing data, whether it comes from Stage 1 (Process Design) or Stage 2 (PPQ).

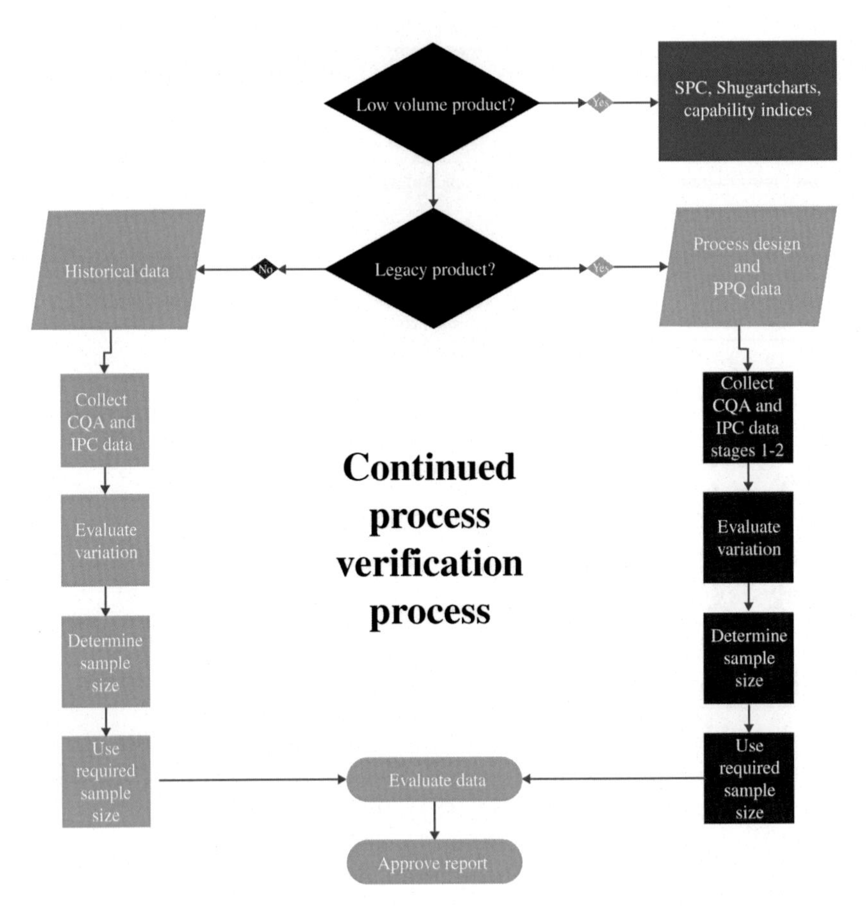

Figure 9.1 Sample CPV process design.

For the creation of a Stage 3a (initial CPV) plan for a new product a data from 20 to 25 batches would be needed.

The following are the points to consider for a new product CPV:

- Creation and execution of a Stage 3a CPV Plan (a prospective plan to assess batches that have not yet been produced) for a minimum of 20–25 batches with a statistically and risk-based justification for the number of samples taken.
- Creation of Statistical Process Control (SPC) Control limits or targets for a Stage 3b CPV plan using data collected and analyzed from the 3a plan.
- Creation of statistically rationalized statistically and risk-based 3b sampling plans for materials attribute (MA), CQA, and IPCs.

- Monthly, quarterly, or yearly review of the capability of the MA, CQA, and IPCs.
- Monthly, quarterly, or yearly review of the MA, CQA, and IPCs control charts.
- Monthly, quarterly, or yearly review of Control strategies for CPPs.

The following are the points to consider for a legacy product CPV:

- Creation and execution of a retrospective 3a CPV Plan for a minimum of 25 batches.
- The plan should include identification of the batches used for the review.
- Creation of SPC Control limits or targets for a 3b CPV plan using data collected and analyzed from the 3a plan.
- Creation of statistically and risk-based rationalized 3b sampling plans for MA, CQA, and IPCs.
- Monthly, quarterly, or yearly review of the capability of the MA, CQA, and IPCs.
- Monthly, quarterly, or yearly review of the MA, CQA, and IPCs control charts.
- Monthly, quarterly, or yearly review of Control strategies for CPPs.

Continued Process Verification Strategy and Enhanced Sampling

The following rules are recommended for Stage 3a CPV sampling for New Products. The rules for a typical Stage 3a CPV process for New Products will apply the following steps and criteria shown in Table 9.2:

- If the PPQ Process Capability for a MA, CQA, or IPC has been calculated to be greater than or equal to 2.0 (as calculated with a 95% confidence interval for Ppk or using 1.5 sigma shift calculations) the attribute can be sampled at its release specification criteria for a minimum of 20–25 batches and a minimum time period of 6–12 months. This provides meaningful statistical sample size and appropriate review of input variability (shifts, incoming material, Active Pharmaceutical Ingredients (API), and personnel).
- If the PPQ (or Legacy review) process capability has been calculated to be less than 2.0, but greater than 1.0 (as calculated with a 95% confidence interval for Ppk, or using 1.5 sigma shift calculations) a

Table 9.2 CPV Stage 3a Sampling and Evaluation

CPV 3a Sampling ($N = 25$ Batches)	Ppk Capability Level With 95% CI or 1.5S Shift (or Equivalent PPM)	Batch Sampling
Risk assessment and Plan Required	Ppk <1.0	Continue with Tolerance Analysis Sampling from PPQ (Enhanced Sampling)
Low to Moderate Risk	$1.0 \leq$ Ppk < 2.0	$N > 1$, with a Rational Sampling Plan
Low Risk	$2.0 \leq$ Ppk	Release sampling plan

rational sampling plan for the attribute at a rate greater than $n = 1$ for a minimum of 20–25 batches and a minimum time period of 6–12 months will be created.

- If the PPQ process capability for a given MA, CQA, or IPC has passed tolerance analysis testing based upon its relative criticality and has a Ppk less than 1.0, a risk assessment should be performed to determine the ability to proceed into commercial manufacturing as well as to identify source of variation. If the overall risk has been determined to be acceptable, an enhanced sampling plan should be created, and attribute specific outputs should be reviewed frequently whether the attribute being reviewed is a part of 3a or 3b. If the attribute's Ppk (or equivalent) increases to above 1.0 for a predetermined time period, then the review period can move to a quarterly basis.

The following are the rules for 3b CPV Sampling and Review for New and Legacy Products. The rules for a typical Stage 3b CPV process will apply the following steps and criteria shown in Table 9.3:

- If the Stage 3a CPV plan's data review (prospective for New Products, retrospective for Legacy) shows that the Process Capability for a MA, CQA, or IPC has been calculated to be greater than or equal to 1.67 (as calculated with a 95% confidence interval for Ppk, or using 1.5 sigma shift calculations) the attribute can be sampled at its release specification criteria during Stage 3b. It should be reviewed periodically or when an OOS occurs that points to the process.
- If the 3a CPV plan data review shows that the process capability for a MA, CQA, or IPC has been calculated to be less than 1.67, but greater than 1.33 (as calculated with a 95% confidence interval for Ppk or using 1.5 sigma shift calculations) the attribute can be sampled at its release specification criteria during Stage 3b and should be reviewed at least quarterly.

Table 9.3 CPV Stage 3b Sampling, Evaluation, and Review Periodicity

CPV 3b Sampling and Review Period	Ppk Capability Level With 95% CI or 1.5S Shift (or Equivalent PPM)	Batch Sampling
Risk assessment and Plan	Ppk < 1.0	Tolerance Analysis (enhanced sampling)
Quarterly	1.0 ≤ Ppk < 1.33	Release with rational sampling (e.g., B, M, E)
	1.33 ≤ Ppk < 1.67	Release sampling plan
Yearly (PQR)	1.67 ≤ Ppk	Release sampling plan

- If the 3a CPV plan data review shows that the process capability for a MA, CQA, or IPC has been calculated to be less than 1.33, but greater than 1.00 (as calculated with a 95% confidence interval for Ppk or using 1.5 sigma shift calculations) the attribute can be sampled at its release specification criteria during Stage 3b and should be reviewed at least quarterly. Note: Care should be taken to apply rational sampling (B, M, E).
- If the 3a CPV plan data review (or Legacy review) shows the Ppk for a MA, CQA, or IPC has been calculated to be less than 1.0 (as calculated with a 95% confidence interval for Ppk, or using 1.5 sigma shift calculations) a risk assessment should be performed to determine the ability to continue in commercial manufacturing as well as to identify source of variation. If the overall risk has been determined to be acceptable, an enhanced sampling plan should be created, and attribute specific outputs should be reviewed at least monthly until the capability reaches a minimum of 1.0.

The following are rules for Low Volume Products. The Drug Substances, API, and Drug Products that will have less than 10–15 batches produced over a 3- to 5-year period (defined by actual legacy batch review demand or future forecasting) may use alternative methods to evaluate capability and interbatch trending. These methods may include parametric and nonparametric methods such as:

Hypotheses Testing

- one-way t tests;
- one-way Z tests;
- two sample t tests;
- ANOVA; and
- standard deviations test.

Nonparametric Testing

- one sample Wilcoxon;
- one sample % Defective;
- two Sample % Defective;
- Mann–Whitney;
- Kruskal–Wallace; and
- Mood's Median.

For the products with expected low volumes the following are points to consider:

- Review existing stage 1 and stage 2 data (or historical data if a legacy product).
- Determine acceptable, statistically derived targets to which each individual 3a batch will be compared. This target may be acceptable difference from target or inter batches difference.
- Determine the sample sizes that will provide the appropriate power for the analysis of each batch to the target as well as for comparison of the entire set of batches (inter and intra batch analysis). A Stage 3a CPV for Low volume products should include appropriately statistically powered number of batches to ensure the initial targets are effective and enough incoming variation is included in the overall analysis.
- After confirming the target, Stage 3b for low volume products may concentrate on comparing the batches to the target using the hypothesis tests shown above.
- Targets should be reviewed at least every 12–24 months.

If, over time, the total number of batches reaches 20–25, it may be appropriate to create control limits and start charting the specific attribute using SPC.

Let us take an example of a product that has met the requirements for a low volume and a CPV plan has been established. A data have been collected on a critical quality attribute which has a release a certain specification. If data are shown to be normally distributed with an historical average of 0.25 of the specification and overall standard deviation of 0.05 of the specification, an example target may be set to be <0.95 of the specification and the tool to be used for comparison is a one-way *t*-test that compares a new data set to an existing target or historical average given an unknown standard deviation.

Maintenance of Validation and Change Control, and Periodic Assessment

As described earlier, the CPV is a stage of Process Validation at which maintenance of the program occurs. Mainly we are in a data collection and trending mode. This is an intensive knowledge accumulation and continued learning phase. Undoubtedly, knowledge of the product critical quality attributes drove the development of the appropriate in-process controls, monitoring parameters and release specifications for the Drug substance and Drug product. Development of control strategies for new products and processes should start with the foundation of knowledge gained from earlier products in order to more efficiently develop in-process controls, monitoring, and release methods—thus reducing the development life cycle for the product. In addition, sharing of best practices for nonproduct specific operations across the manufacturing organization, such as equipment cleaning, calibration, environmental monitoring, and control also increase efficiency and help ensure product safety, as well as provide additional information on variability of the process. Conversely additional controls may be added as a result of the better understanding of quality attributes that must be tracked and monitored, refined analytical methods, evolving sample/testing strategies, long pool hold times, and other factors related to either product quality measurement or plant constraints. Retention of laboratory data, as well as clinical manufacturing in-process control, monitoring, and release data is required for CPV and product commercialization. Learning from Process monitoring, deviations/nonconformances investigations, etc. are regulatory expectations. Additionally, formal lessons learned activities at key milestones in development can contribute to the total base of organizational knowledge and process understanding for future programs. As more manufacturing lots are produced, statistical control strategies are put in place to help monitor process consistency. This knowledge can drive continuous process improvement, as well as identify trends that point to possible areas which need greater operational control. As process understanding increases, certain in-process controls and monitoring strategies may be updated for process steps that are consistently robust at commercial scale. Conversely, additional controls may be put in place, or control limits tightened in those instances where the perceived criticality of certain quality attributes increases, or process robustness is not as great as expected. Continuous process improvements are driven by greater product knowledge as more patients are treated with the drug, as well as by process monitoring and other tools that are used to gain better understanding of the process a commercial scale. Finally, product and process knowledge from commercial and later stage development products is being fed back to development scientists to identify products with superior quality attributes and develop more robust and/or streamlined processes.

Appropriate change control system which should be a part of the Pharmaceutical Quality Systems must work hand in hand with CPV and be agile to accommodate possible future changes, as level of product and process knowledge increases. It is recommended to review CPV program on at least a quarterly basis if it is not a low volume product. There are some indications from currently interviewed industry QA professionals (based on their FDA inspections' experiences) that FDA expects at least this periodicity for a typical non–low volume product review.

Develop Control Rules for Continued Process Verification

The following rules are recommended for Control Charts to promote consistency of CPV review data. It should be noted that selecting the proper control chart for use in the CPV plan requires an understanding of the following:

- The type of data being collected, discrete versus continuous.
- How the data are being collected.
- How many samples are being taken at a given time.

The purposes for the use of control charts include:

- To monitor and provide warning for processes going out of control.
- To keep from making process adjustments when they may not be needed and adjusting when only when required.
- To determine the natural range of the process and to compare this range to its specification limits.
- To inform about the process' capability and stability.

There are two main families of control charts based upon whether the data collected are continuous or discrete. The control charts discussed in this chapter are univariate and either monitor attribute or variable data, as the monitoring program becomes more mature in the use of these tools, more complicated charting tools (multivariate) may be used. The control charts for continuous data are also called variables control charts and should be used under the following conditions:

- If the data are collected in subgroups and the number of samples in the subgroup is less than 8, the $\overline{X}$ (Average)–R (Range) charts should be used.
- If the data are collected in subgroups and the number of samples in the subgroup is greater than 8, the $\overline{X}$ (Average)–S (Standard Deviation).
- If the data collected are a singular sample, the I/MR (individuals and moving range).

If the data collected are discrete, then the attributes family of control charts can be used. The type of chart is dependent upon how the defectives are going to be reported.

- If the data are going to be reported in proportion defectives, the P-chart (proportion of nonconforming units) is used.
- If the data are reported in the number of defective units per subgroup, the nonparametric (NP) chart number of nonconforming units is used.
- If the data are going to be reported as the number of defects per unit sample, the U-char (count type data) is used.
- If the data are going to be reported as the number of defects in the subgroup, the C-chart (count type data) is used.

Trending analysis is aimed at knowing the common cause process variation and detecting special cause variation within each batch. For each attribute being trended, evaluate the data set to identify excursions, shifts, and trends. The following trending analyses test and rules are recommended for data evaluation:

- Excursions are identified by results that fall outside of the calculated 3σ control limits of the data mean.
- Shifts are identified if there are nine or more consecutive point on the same side of the data average.
- Trends are identified if there are six or more consecutive point increasing or decreasing.

If any excursions, shifts, or trends are identified, an additional assessment is required to identify a special cause. Typical investigation would include review of deviations for the batch, review of batch record parameters any comments, change to a new raw material lot, and/or any change controls for equipment or process changes.

When outliers due to special causes are identified, they may be eliminated from the statistical evaluation, so that future data are evaluated against limits derived only from common cause variability. Only for cause elimination should occur as there is a risk removing data when, in fact, it should be included in the data set. This may require multiple iterations of control charts to be created and evaluated before finalizing a CPV report.

This trending must be reviewed per the period defined in the Stage 3b CPV plan for the product and process. Investigations must be tracked and assessed within the sites quality management systems. Limits and subsequent review periods may be updated based upon the results of the investigation.

Process capability should be measured in terms of actual capability (Ppk). It is suggested to present Ppk with 95% confidence intervals. Actual process capability (Ppk) is a measure of a process central tendency and variation relative to specification. Ppk is dependent on the assumption of normality. Departures from normality will invalidate conclusions drawn from this statistic.

Ppk = Process Performance index for two-sided specification limit accounting for process centering

$\overline{X}$—process average;

USL—upper specification limit;

LSL—lower specification limit;

σ—standard deviation $\sqrt{\sum_{i-1}^{n} \frac{(x_i-\mu)^2}{n-1}}$

$$P_pK = \min\left(\frac{USL-\mu}{3\sigma} \text{or} \frac{\mu-LSL}{3\sigma}\right)$$

Develop Strategies for Continued Process Verification

CPV procedure must be established within document control system. It is recommended that all CPV plans must be approved and stored within the system. All CPV reviews must be documented and approved. Standard templates are strongly suggested for that purpose. All CPV continuous improvement activities must me tracked and trended within the appropriate site quality management system. There are many factors to be considered when creating a CPV Plan and at a minimum will include:

- The CPV Plan will be reviewed and approved by senior management.
- The CPV plan needs to include a frequency of review of the information from data collection mechanisms as well as Quality Systems.
- It should also identify circumstances for, and a process to allow for, an immediate review based on significant issues identified with a process or product and identify the participants in the review.
- The plan shall address new or legacy product or process.
- The frequency of data review for a CPV is risk-based and will be defined in the plan.
- Rules must be created for analyzing data based upon the statistical tools that are being used.
- A risk-based assessment is used to evaluate excursions, shifts, or trends identified within a control chart. The type and level of investigation executed shall be risk-based and documented in the quality system as required.
- Limits and subsequent review periods may be updated based upon the results of the investigation.

The difference between CPV and Annual Product Review (APR) should also be identified as a part of CPV strategy. It should be noted that CPV plans and reports do not take the place of APR or Product Quality Reviews (PQR). The APR is a review that is annual filed with parenteral product application (NDA, ANDA, BLA, etc.) while CPV is internal data collection, analysis, and evaluation to confirm process consistency and to determine if there any shifts in the process. The APR should include a high-level reviews and summaries of CPV data reviews. The APR should also identify gaps in the CPV data reviews and summarize long-term trends and other activities not captured in the CPV. The CPV and APR efforts should be coordinated to eliminate redundancies and assure proper data collection and flow. CPV data should be monitored continuously and reported on at least quarterly basis at a minimum for not low volume products. The Stage 3 Deliverables may include:

- CPV Plans (3a and 3b as required);
- Quarterly CPV plan reviews and reports;

- Summary CPV reports for APR/PQR; and
- Updated Risk Assessments.

Examples of Case Study Evaluations for Parenteral Products

The following are few examples of data analyses used for CPV. One of the first methods to be considered for analysis should be tolerance intervals.[12] This statistical tool shows a coverage probability as a proportion of items in a population lying within a statistical tolerance interval. In other words, the probability of the statistically significant proportion of population (such as 95%) meeting its tolerance interval based on the number of samples taken (where multiplier of *s* is dependent in a number of samples taken). Fig. 9.2 shows results of Tolerance intervals calculation for two tail population of hypothetical Assay results for at least 14 batches of product. As illustrated in this figure 95% of sampled product population of batches tested will meet specification within its limits with 95% probability, as the lower tolerance interval is at 97% and the upper is at 104%. Let us explore another statistic—process capability for another hypothetical example of % Assay calculated for 28 parenteral product batches and shown in Fig. 9.3. This analysis reveals that process capability indices are very high—9.34. This is well within desired Ppk = 2.00. In addition, we can also see that with 95% probability we can be sure that there will be no failure (occurrences or PPM = 0.00). This well controlled process. Finally we would like to recommend a use of dashboards for ease of review and communication of CPV programs. Fig. 9.4 illustrates an example dashboard. It is conveniently divided into three sections:

- list of the Process Capability Indices calculations for each CQA;
- pie chart showing current state of CQA's capabilities; and
- legend showing firms conventions for Low/Medium/High Risk CQAs' Capability Measurements.

This dashboard easily communicates to reviewing team and to management that this product has an issue with a CQA 5 and possibly CQA 4. Upon review team should review Process Control Strategy and evaluate what improvement could be made to CPPs that control these CQAs in "yellow" and "red."

[12] ISO 16269 Statistical interpretation of data—Part 6: Determination of statistical tolerance.

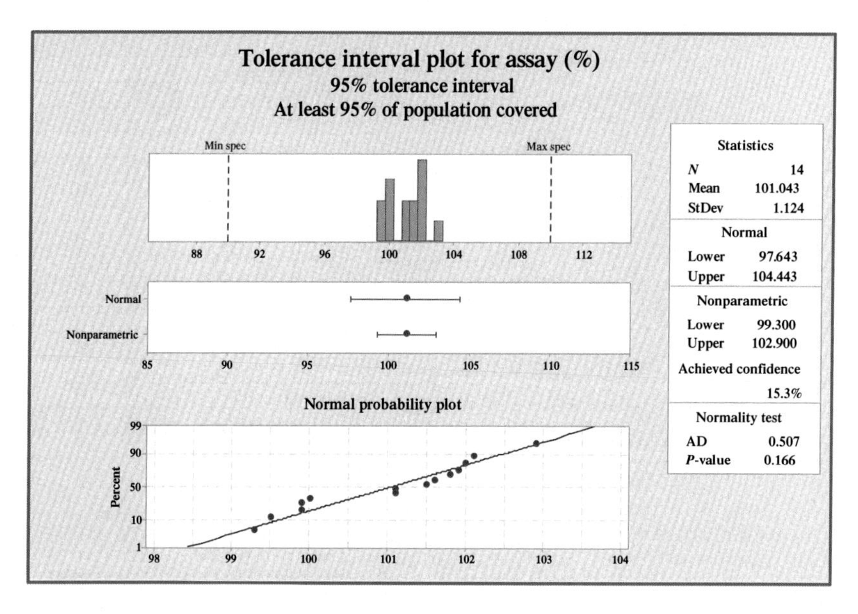

Figure 9.2 Example tolerance chart for a CQA.

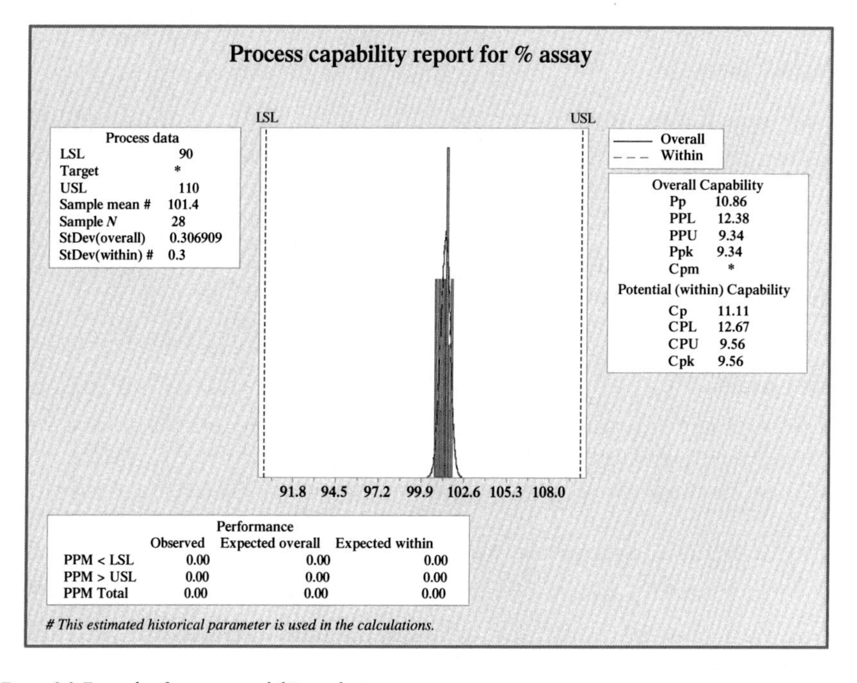

Figure 9.3 Example of process capability analysis.

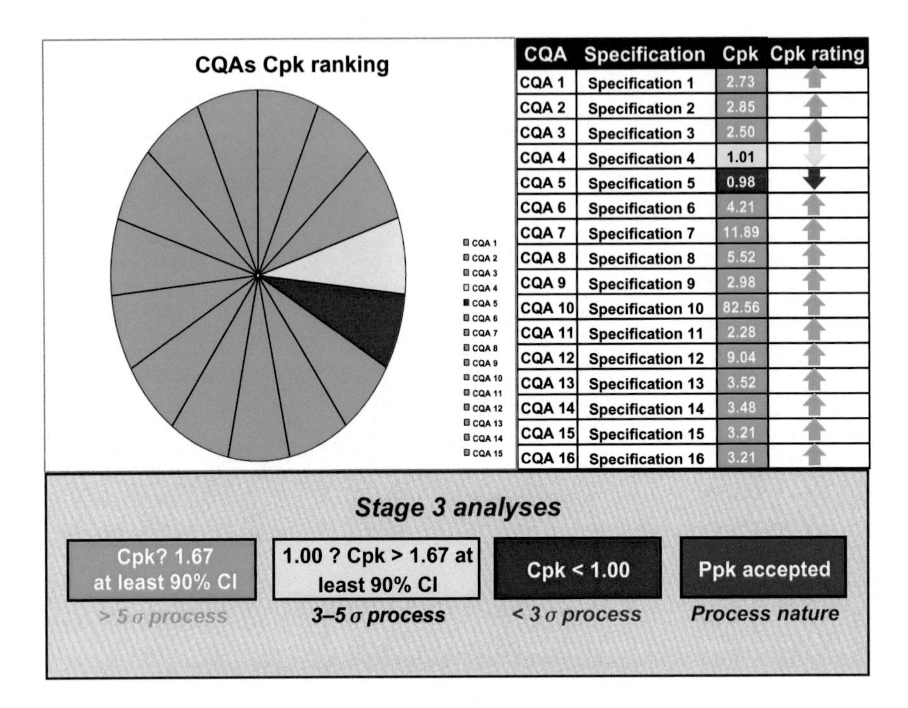

CQA	Specification	Cpk	Cpk rating
CQA 1	Specification 1	2.73	↑
CQA 2	Specification 2	2.85	↑
CQA 3	Specification 3	2.50	↑
CQA 4	Specification 4	1.01	↓
CQA 5	Specification 5	0.98	↓
CQA 6	Specification 6	4.21	↑
CQA 7	Specification 7	11.89	↑
CQA 8	Specification 8	5.52	↑
CQA 9	Specification 9	2.98	↑
CQA 10	Specification 10	82.56	↑
CQA 11	Specification 11	2.28	↑
CQA 12	Specification 12	9.04	↑
CQA 13	Specification 13	3.52	↑
CQA 14	Specification 14	3.48	↑
CQA 15	Specification 15	3.21	↑
CQA 16	Specification 16	3.21	↑

Figure 9.4 An example CPV dashboard.

Conclusion

At the conclusion it is imperative that these data and information including CPV dashboard at firms' Quality Councils (FDA, 2016; Modernizing Pharmaceutical, 2018). In addition, CPV trending should become part of firms' quality metrics. These initiatives will help firms installing and nourishing cultures of science and risk-based approaches enabled by knowledge management.

References

Code of Federal Regulations Title 21--Food and Drugs Chapter I - Food and Drug Administration Department of Health and Human Services Subchapter C--Drugs: General.

FDA, 2004. Guidance for Industry PAT—A Framework for Innovative Pharmaceutical Development, Manufacturing, and Quality Assurance, September.

FDA, 2011. Guidance for Industry Process Validation Guidance: Principles and Practices, January.

FDA, 2016. Submission of Quality Metrics Data Guidance for Industry, Draft Guidance, November.

ICH, 2009. Q8(R2) Pharmaceutical Development, August.

Modernizing Pharmaceutical Quality Systems; Studying Quality Metrics and Quality Culture; Quality Metrics Feedback Program, 06/29/2018. <https://www.federalregister.gov/documents/2018/06/29/2018-14005/modernizing-pharmaceutical-quality-systems-studying-quality-metrics-and-quality-culture-quality>.

CHAPTER 10

Preuse/Poststerilization Integrity Testing of Sterilizing Grade Filter

Maik W. Jornitz
BioProcess Resources LLC, Manorville, NY, United States

Introduction

Sterilizing grade filtration has been reliably used for decades and is becoming increasingly the method of choice to cold sterilize fluids and gases. The multitude of launched biologic drug products can only be sterilized by filtration, even when heat sterilization may be the preferred method, one should embrace the fact that filtration is a robust form of sterilization. The reliability of the filtrative aseptic processing step increased with filter stability improvements, robust integrity test methodologies, and especially process validation requirements, which evaluate the performance of a sterilizing grade filter under process conditions utilizing either the actual product or a placebo, if the fluid is bactericidal or bacteriostatic (PDA Technical Report 26, 2008; Food and Drug Administration (FDA), 2004; ISO 13408-2:2003(E), 2003).

To routinely verify that the sterilizing grade membrane filter is flawless, nondestructive integrity tests, like Bubble Point, diffusive flow, or pressure decay, are used successfully since the first regulatory requests in the mid-1970s. The integrity test can be performed preuse/presterilization, preuse/poststerilization and must be performed postuse; use meaning the actual filtration process. The proportion of the actual integrity test employment varied in the past, as an integrity test is a filter manufacturer's release criteria and the transport packaging of such filters is thoroughly tested (Table 10.1). With that preuse tests may seem unnecessary, if not risk elevating when the filtrative assembly is presterilized.

The integrity test of a sterilizing grade filter has to be and most commonly is performed after the filtration process (postuse).

Principles of Parenteral Solution Validation. DOI: https://doi.org/10.1016/B978-0-12-809412-9.00009-5

Table 10.1 Estimation of the Integrity Test Employment Ratio Before Enforcement

	Preuse	Preuse	Postuse
	Presterilization	Poststerilization	
Employment ratio	~10%–20%	<5%	100%

Some filter users test the integrity before the filtration process and before the filter is sterilized. Exceptionally rare are integrity tests preuse/poststerilization, as such test would require downstream, filtrate side manipulation, and therefore is considered precarious to the sterilized filtrate side.

Regulators request the test of the integrity of a sterilizing grade filter after the filtration, postuse, but propose an integrity test preuse, without specifying nor enforcing whether pre- or poststerilization (Food and Drug Administration (FDA), 2004; ISO 13408-2:2003(E), 2003; Ministry of Health, 2005; Pharmaceutical Inspection, 2004; World Health Organization (WHO), 2009). However, EU Annex 1 (EudraLex, 2008), paragraph 113. states: "113. The integrity of the sterilised filter should be verified before use and should be confirmed immediately after use by an appropriate method such as bubble point, diffusive flow or pressure hold test..." The paragraph, in its first sentence, endorses the rare and risk attached preuse/poststerilization integrity test, which is, questionably necessary. This in itself is not new, as this statement has been made in earlier versions of Annex 1; however, it seems that the word "should" is nowadays enforced as "must" by European inspectors. This causes major problems within the industry in new and existing applications. Where other regulatory authorities keep the choice to the filter user and process accountable, Annex 1, paragraph 113 implies enforcement or at least cause misunderstanding of the meaning of "should." One has to investigate the usefulness if not risks of preuse/poststerilization integrity testing.

Possible Reasons to Apply Preuse/Poststerilization Integrity Testing

As mentioned, preuse/poststerilization integrity testing has not been common practice, but rather a rare event, knowing that such test would require a downstream manipulation. This means that the vast majority of sterilizing grade filtration application perform very successful, processing a sterile filtrate. Incidences of postuse filter failure, another rare occurrence, were detected and the filtered product either discarded or reprocessed. So, the question prevails, why one would run a preuse/poststerilization test. The only reason why one would enforce a preuse/poststerilization integrity test is the notion that the postuse integrity test would not be able to detect a filter failure due to the blocking of the filter flaw.

Sterilizing grade filtration systems require to be sterilized; either presterilized, for example, by gamma irradiation, when assembled to a containment single-use hold bag or sterilized at the site after assembly into the process or filter housing. The presterilization, gamma irradiation process is highly qualified by the vendor of the single-use assembly and experience show that filter systems are not at risk to be damaged by this process. However, when the filter is installed into a stainless-steel filter housing and in situ steam sterilized, filters experience thermal and mechanical stresses. Hence steam sterilization qualification is a necessity not only to determine the sterilization efficacy, but also whether the sterilization process stays within the maximum operating parameters given by the filter manufacturer and the filter is not damaged by the sterilization cycle. If the qualification process has either been insufficient or an end user mistake occurs, the filter may be damaged during steam sterilization, which can result in a retentively underperforming filter. The apprehension in this case is twofold: (1) the pore structure or parts of it become enlarged due to the steam sterilization and (2) a minor flaw is created. This in itself may cause the problem of a potential unsterile filtrate, but normally the flawed filter is detectable by the postuse integrity test and necessary activities can be taken to avoid the release of the filtered product. Though the European regulatory hypothesis is that in either instances, the enlarged pore structure or minor flaw are plugged by contaminants separated from the fluid and are not detected by the postuse integrity test (EU GMP Guide, 2007). Based on these two theories, the preuse/poststerilization integrity necessity has been enforced.

Albeit the hypothesis that an enlarged pore structure or minor flaw within a pleated sterilizing grade filter will be covered by contaminants so that the postuse integrity test cannot detect it, has not been experienced, according to discussions by filter manufacturers (Thomas, 2011; PDA, 2012). The manufacturers have performed thousands of bacteria challenge tests, subjecting filter units to 10^7 *Brevundimonas diminuta* per square centimeter, which represents a very high contamination level in the feed stream (ASTM, Standard F838-15, 1983). Typically the integrity of the challenged filter is tested before the bacteria challenge and postchallenge. Results showed that filters, which failed the integrity test before being subjected to the high bioburden challenge test have also failed the integrity test postchallenge. The evaluation showed that the postuse integrity test is sensitive enough to detect any flawed filter, even when exposed to a severe contamination level. It has to be pointed out that one typically does not find such high contamination levels within routine filtrative sterilization applications.

To further confirm the stated and gain additional assurance that the postuse integrity test does expose a flawed filter, no matter what blockage of the membrane, tests were performed at various blocking rates with 0.2 micron filters, some which just failed or passed the preuse integrity test (Jornitz and Stering, 2017). The data showed that the postuse test would detect a flawed filter, independent of the contamination rate.

The Risks Attached to Preuse/Poststerilization Integrity Testing

When one appraises the different integrity test methods, one realizes that any integrity test requires manipulation of the downstream side, as the filter needs to be wetted with fluid, except when performing the postuse test, a product wet integrity test could be performed. In addition, when the integrity test is performed, the downstream, filtrate side requires being under atmospheric pressure. Both wetting and venting manipulations are most commonly undesirable after the filter and filtrate side have been sterilized, either in situ by steam or as single-use assembly by gamma irradiation. The severity of a sterile filtrate side manipulation can be as critically compared as any human intervention in a filling process. In both instances, the sterile side of the process is potentially subjected to environmental influences with the risk of microbial ingress.

Certainly engineering options are or can be made available to divert the wetting fluid into a fluid receiving tank or container, as well as utilize this receiver or other options for venting purposes (Fig. 10.1).

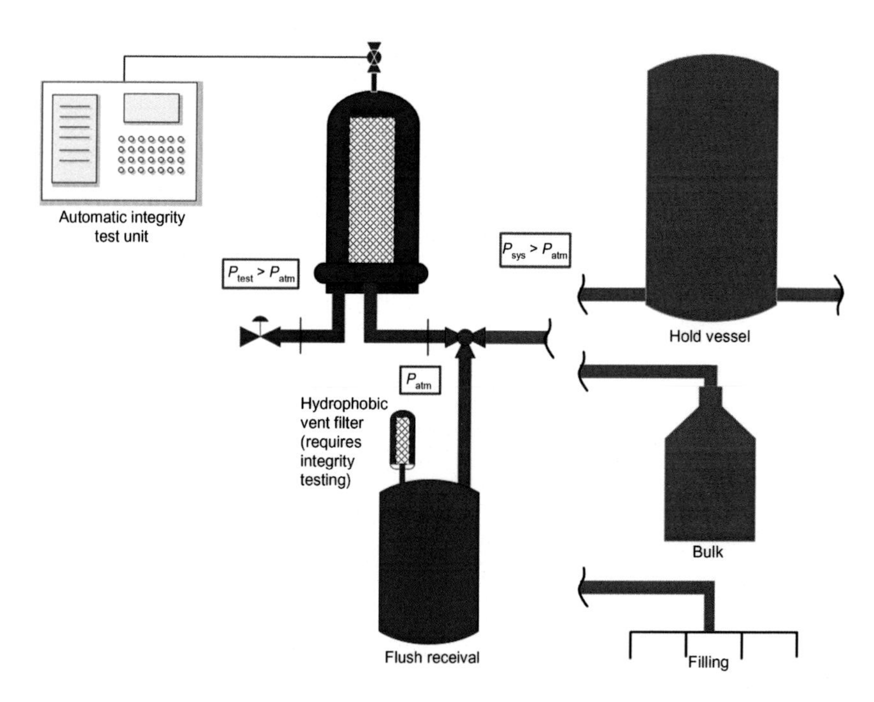

Figure 10.1 Integrity test and downstream setup example for preuse/poststerilization integrity testing.

Nevertheless any supplementary design components create an increase in connections, risk and potential of oversight, or mistakes. Vent filters or flush filter barriers require integrity testing and the associated tasks, installation, steaming, and flushing, drying. Any added equipment and piping need to be cleaned, setup, and sterilized. All gaskets require inspection and potential testing. To summarize, there are engineering solutions possible and accomplishable; however, it creates an additional burden and risk to an already complex downstream process (Fig. 10.2). An increase in the filtration system complexity furthers human error mistakes or equipment failure. The desired design of such systems requires to be simplistic and robust. Robustness of a process is not getting enhanced with a multitude of downstream connections or manipulation of the sterilized filtrate side.

More importantly, sterilized filtrate process equipment is commonly kept under pressure after steam sterilization or in case of single-use systems are preassembled containment systems. The introduction of a vent filter and/or atmospheric pressures on the sterilized downstream side can cause an undetected breach of the sterile filtrate side. Preuse/poststerilization integrity test enforced to presumably increases the quality of a sterile drug

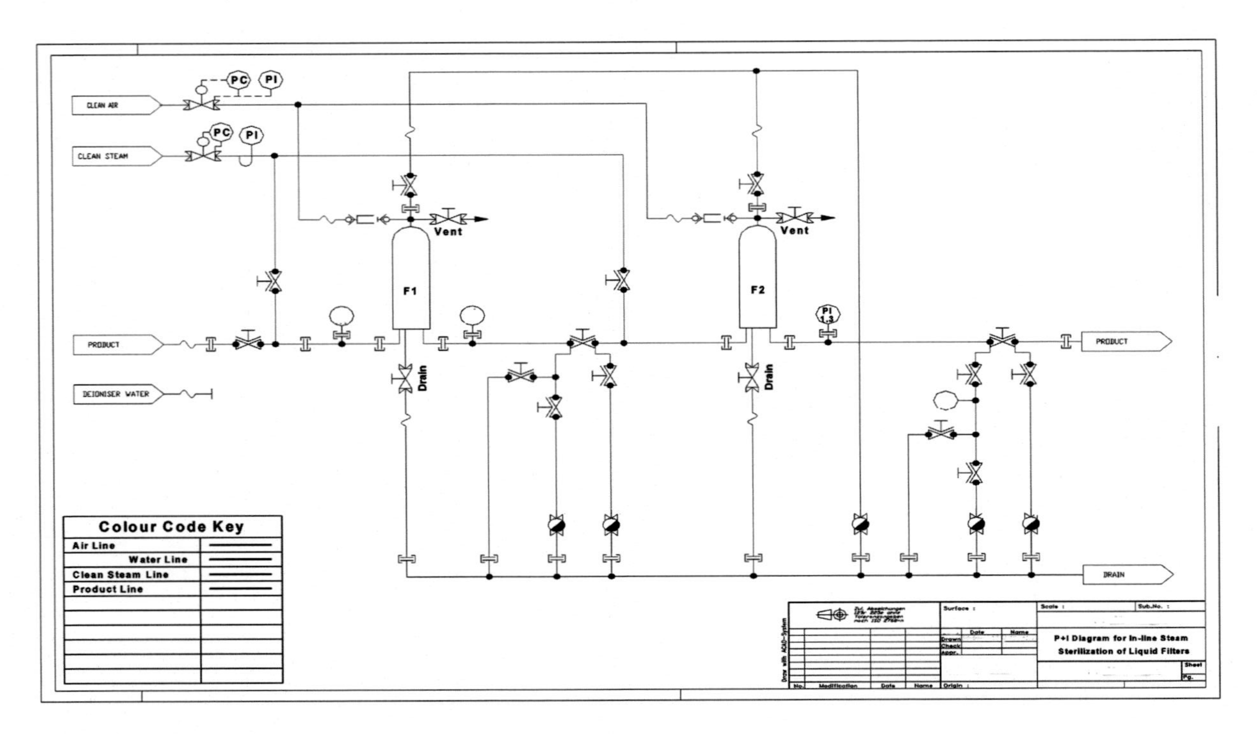

Figure 10.2 Example of a filter setup with associated pipe work.

Table 10.2 Severity–Frequency–Detectability for Filtration Risk Assessment

Value	Severity
1	Negligible: Has no potential to have an adverse effect on identity, strength, quality, purity, or potency of a drug product
2	Minor: Has minimal potential to have an adverse effect on identity, strength, quality, purity, or potency of a drug product
3	Moderate: Has moderate potential to have an adverse effect on identity, strength, quality, purity, or potency of a drug product
4	Major: Has a substantial potential to have an adverse effect on identity, strength, quality, purity, or potency of a drug product
Value	**Frequency**
1	Highly unlikely: The probability of the event occurring is so low that it can be assumed that the event will not occur
2	Unlikely: Event not expected to occur, but theoretically possible
3	Likely: Event may occur and/or has occurred in the past
4	Highly likely: Event expected to occur
Value	**Detectability**
1	Readily detectable: Will be detected
2	May be detectable: May be detected
3	Not detectable: No mechanism for detection

product, the safety towards the patient, represent now a risk elevation as the sterile filtrate side is manipulated with. Utilizing the PQRI (2007) risk assessment process (similarly as discussed in Chapter 3: Quality Risk Management), the actual risk of a preuse/poststerilization activity is higher than the absence of this activity.

According to this document the level of risk is calculated as:

$$\text{Risk} = (S) \times (F) \times (D)$$

where (S) is the severity of the event (consequence); (F) is the frequency estimation (likelihood of event occurring); and (D) is the level of detectability.

The three components for Filter Risk Assessment are measured as shown in Table 10.2.

The scenario of the use of the preuse/poststerilization integrity test and the lack thereof are now compared side-by-side as shown in Table 10.3.

The above-listed risk assessment is an example and shows that the risk using the preuse/poststerilization integrity test is increasing the risk. Other risk assessments by filter users showed similar results.

Table 10.3 Preuse/Poststerilization Integrity Example

Category	w/ Test	w/o Test	Rational
Severity	4	4	If the filter fails or microbial ingress happens, it has a major effect in both cases
Frequency	3	1	Microbial ingress into the downstream side may occur, for this reason the downstream side is commonly held under pressure after steam sterilization. This overpressure is replaced by atmospheric pressure during the preuse test (3)
			Since steam sterilized downstream processes are commonly held at overpressure the likelihood of ingress is low (1)
Detectability	3	1	If there is a microbial ingress due to downstream manipulation, when the test is performed, it will not be detected (3)
			If the filter is flawed due to the steam sterilization process it will be readily detected by the postuse test (1)
Risk level	36	4	Risk assessment would not recommend a preuse/poststerilization test

The lack of detection of a potential breach or an ingress or mistake made elevates the risk level of performing the preuse/poststerilization test. The benefit of such test is questionable, this much, that it should be left to the end user to choose whether or not to perform such test. Regulatory enforcement supposedly increases the patient safety and enhances the robustness and quality of a process. In the case of the preuse/poststerilization integrity test such commendable desire is voided and higher risks are exhorted on current and future filtration processes.

Other aspects to consider, before one chooses to perform a preuse/poststerilization test:

- Every single sterilizing grade filter is integrity tested by the filter manufacturer before it leaves the facility.
- Packaging and transportation of filters is validated by drop, shake, and vibration tests.
- Fluid streams used within the industry are commonly not as contaminated as experienced during bacteria challenge tests with 10^7 cfu/cm^2 (ASTM, Standard F838-15, 1983). However, in these tests it has never been experienced that a preuse integrity tested failed filter became integral and passed the postuse test.
- Minor flaws could be 0.45 μm, which often create retention of 10^5 cm^{-2}, depending on the process conditions.
- Steam sterilization damaged filters are regularly dramatic failures (Fig. 10.3).

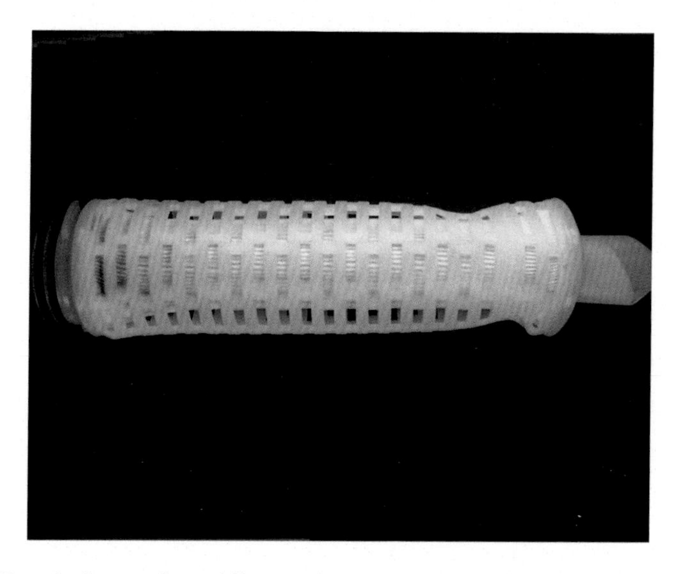

Figure 10.3 Example of a steam damaged filter cartridge. Courtesy Sartorius Stedim Biotech GmbH.

- Postuse integrity test failures due to wrongly installed or damaged filters are an economical risk, not a drug safety risk, as the drug is either discarded or reprocessed.

Recommendations

The recommendations resulting from the analysis of the pros and cons of preuse/poststerilization integrity test are either reword paragraph 113. "The integrity of the *sterilising grade* filter *may* be verified before use and should be confirmed immediately after use by an appropriate method such as bubble point, diffusive flow or pressure hold..." Or, the other option would be to utilize the approach of the FDA (ICH, 2009), which defines the word "should" as a suggestion or recommendation, leaving the decision to the end user in the first place. Since postuse testing is of importance to confirm the filters separation performance during the process, the paragraph might be changed to "The integrity of the sterilised filter should be verified before use and *must* be confirmed immediately after use by an appropriate method such as bubble point, diffusive flow or pressure hold..."

Both alternatives leave the decision to the filter user, who has the unquestionable desire to produce and deliver a safe drug product. The filter users risk assessment requires to be the deciding factor whether or not a preuse/poststerilization integrity test is performed. Generic enforcement does not support the desired state of increasing the patient's safety or risk reduction, but rather increases complexity and hazards.

As many times, stated "you cannot test quality into your product, you have to produce it."

References

ASTM, Standard F838-15, 1983. Standard Test Method for Determining Bacterial Retention of Membrane Filters Utilized for Liquid Filtration. American Society for Testing and Materials, West Conshohocken, PA, Revised 1988, 2005, 2015.

EudraLex, 2008. Volume 4, EU Guidelines to Good Manufacturing Practice Medicinal Products for Human and Veterinary Use, Annex 1, Manufacture of Sterile Medicinal Products, Brussels.

EU GMP Guide, 2007. Annexes - Supplementary Requirements - Annex 1 Manufacture of Sterile Medicinal Products 1. Question (H + V June 2007): How should the integrity of sterilising filters be verified?

Food and Drug Administration (FDA), 2004. Guideline on Sterile Drug Products Produced by Aseptic Processing. Division of Manufacturing and Product Quality, Office of Compliance, Center for Drugs and Biologics, Rockville, MD.

ICH. Q8, Q9, Q10 Guidance for the Industry, FDA, Rockville, MD, 2006, 2006, 2009.

ISO 13408-2:2003(E), 2003. Aseptic Processing of Health Care Products – Part 2: Filtration. ISO Copyright Office, Geneva.

Jornitz, M.W., Stering, M., 2017. Pre-use/post-sterilization integrity testing of sterilizing grade filter – is the post-use test sufficient? Am. Pharm. Rev.

Ministry of Health, 2005. Labour and Welfare (MHLW), Sterile Drug Products Produced by Aseptic Processing. Tokyo.

PDA, 2012. IG12 Filtration Interest Group Presentation by Millipore, Pall and Sartorius, PDA Annual Meeting Phoenix, April 16–18.

PDA Technical Report 26, 2008. Liquid Sterilizing Filtration. Parenteral Drug Association, Bethesda, MD.

Pharmaceutical Inspection, 2004. Convention (PIC/S), Recommendation on the Validation of Aseptic Processes, Geneva, PI 007-2.

PQRI, 2007. Post Approval Changes for Sterile Products Working Group.

Thomas, P., 2011. May Filtration: Debating Post-Sterilization Testing. Pharmaceutical Manufacturing.

World Health Organization (WHO), 2009. WHO Good Manufacturing Practices for Sterile Pharmaceutical Products, Geneva, QAS/09.925 Rev1.

CHAPTER 11

Environmental Monitoring

Igor Gorsky
Senior Consultant, ConcordiaValsource LLC., Downingtown, PA, United States

(As stated in EMA Annex 1 (EudraLex, 2008) "clean areas for the manufacture of sterile products" must be "classified according to the required characteristics of the environment. Each manufacturing operation requires an appropriate environmental cleanliness level in the operational state" to "minimize the risks of particulate or microbial contamination of the product or materials being handled." Mainly this excerpt talks regarding three subjects:

- appropriate design of rooms and facility;
- appropriate selection of filters and determination of airflow to maintain certain level of room pressurization; and
- appropriate monitoring program established to measure, analyze, evaluate and maintain appropriate room classification.

There are a great number of standards[1] (International Standard Organization; International Standard Organization, 2015) and guidances (FDA, 2004; EudraLex, 2008) that provide needed information regarding expected requirements for appropriate environmental conditions for a suitable level of aseptic processing (based on process) that is necessary for most parenteral products manufacturing. Therefore we will provide some examples of means of Environmental Monitoring (EM) Program including personnel monitoring, while performing risk and science-based analysis using EM Risk Evaluation Method (EM-REM) (Stabler and Famili, 2015) or equivalent risk-based analysis. We will also suggest recommendations for improvements of the EM to comply with a current revision of ISO 14644-1, Second edition 12-15-2015 Cleanrooms and Associated Controlled Environments—Part 1: Classification of Air

[1] International Standard Organization, Cleanrooms and Associated Controlled Environments, Part 2, Specifications for Testing and Monitoring to Prove Continued Compliance with ISO 14644-1, International Standard ISO 14644-2 (2015).

Principles of Parenteral Solution Validation. DOI: https://doi.org/10.1016/B978-0-12-809412-9.00010-1

Cleanliness by Particle Concentration. Already mentioned earlier EMA Annex 1 also states that "clean rooms and clean air devices should be routinely monitored in operation and the monitoring locations based on a formal risk analysis study and the results obtained during the classification of rooms and/or clean air devices." That is a very important recommendation. Using Quality Risk Management should be in a core of the properly established EM Program.

A typical pharmaceutical or biotechnology company developing and producing parenteral products must design their clean rooms and associated controlled environments to provide for the control of contamination of air and, if appropriate, surfaces, to levels applicable for accomplishing contamination-sensitive activities. Contamination control is beneficial for protection of product and process integrity in parenteral products manufacturing. Recent revision of ISO 14644 specifies classes of air cleanliness in terms of the number of particles expressed as a concentration in air volume. It also specifies the standard method of testing to determine cleanliness class, including selection of sampling locations.

The most significant change in this version of an ISO standard was the adoption of a more consistent statistical approach to the selection and the number of sampling locations; and the evaluation of the data collected. The statistical model is based on adaptation of the hypergeometric sampling model technique, where samples are drawn randomly without replacement from a finite population. The new approach allows each location to be treated independently with at least a 95% level of confidence that at least 90% of the cleanroom or clean zone areas will comply with the maximum particle concentration limit for the target class of air cleanliness. As per this new version of ISO, the cleanroom or clean zone area is divided up into a grid of sections of near equal area, whose number is equal to the number of sampling locations derived from Table A.1 in the standard. A sampling location is placed within each grid section to be representative of that grid section.

It is assumed for practical purposes that the locations are chosen representatively; a "representative" location means that features such as cleanroom or clean zone layout, equipment disposition and airflow systems should be considered when selecting sampling locations. Additional sampling locations may be added to the minimum number of sampling locations.

Therefore an aim of manufacturing EM strategy is to align risk factors and risk-based approach for EM sampling strategy. We will discuss an example of legacy program. First step in this kind of exercise is to analyze data available from current EM routine studies to establish those areas that are most prone to probability of not meeting ISO 14644-1 specifications.

Review revealed that all of the steps of the process are carefully designed to optimize reduction of cross-contamination from process variables, such as personnel, environment, facilities, equipment, and process steps using models similar to EM-REM such as optimization of proximity of the sampling locations to the process, as well as optimization of a number of an essential personnel being performing the process in the room. Review of the EM Program and interviews with personnel may also help to show the how well the Process Quality at the firm is designed into an Aseptic Processing by:

- Aseptic Process Simulation Program;
- Aseptic Process Validation (to be performed based on Risk Management Program);
- Aseptic Operator Qualification;
- Integration/Intertwined Many Aseptic Controls;
- Clean Room & Hoods Design and Engineering;
- Risk-Based Aseptic Operations Personnel Monitoring; and
- Risk-Based Clean Room Environmental (& Personnel) Monitoring.

In addition, one should also investigate incidents of bioburden results not meeting specifications. A well-designed facility with a well-designed maintenance and monitoring plan should have very few incidents of microbial failures. These seldom aberrant results should always be identified and corrected.

Let us briefly discuss basic concepts behind EM-REM[2] (Moldenhauer and Madsen; PDA) developed by Valsource team (H. Baseman, M. Hardiman-Steed, W. Henkels, C. Hunff, and Dr. M. Long). The process of developing a program consists of the following:

- Identify the manufacturing process/process mapping [Gemba (Imai, 1997)].
- Perform a risk assessment [typically Ishikawa (Ishikawa, 1976)/ Fishbone diagram] is used for this purpose.

[2] EM-REM is Environmental Monitoring Risk Evaluation Method.

- Assess planned personnel flow.
- Assess planned material flow.
- Sample site locations should be based on risk of activity.
- Determine the duration of each activity.
- Determine the proximity of the operators to the actual process.
- Identify the number of operators in the area.

Than the risk analysis is based on the following Risk Boxes shown in Figs. 11.1 and 11.2.

As shown above this is quite a structured and definite process one could employ, and it clearly demystifies risk management application of the EM-REM. But EM-REM is mainly useful for determination of sampling locations. Further examples of data evaluation are presented below.

The typical evaluation for an example aseptic facility shall be concentrated on analyses of viables' (microbial) data and data from 0.5 μm nonviable particles results from all major processing rooms which were inessential in manufacturing of the product. This particle size is typically chosen as it is found to be most variable as well is for a reason of performing an objective comparison using one factor across entire production flow.

Additionally, as prescribed by ISO 14644-1, each room should be divided into a grid with approximately similar size area quadrants and statistically derived sampling number requirements were applied to each major manufacturing process to implement for a future monitoring.

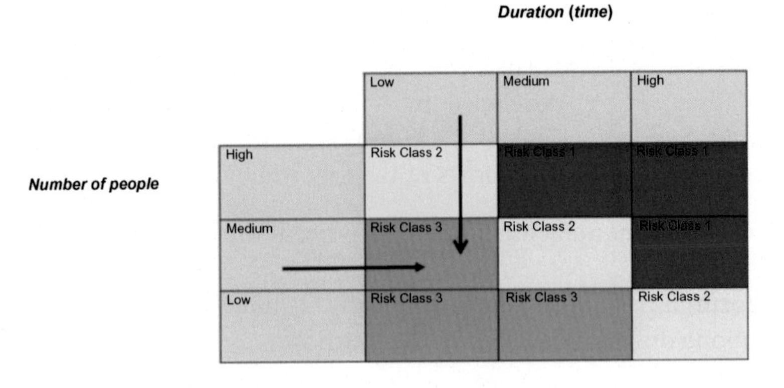

Figure 11.1 An example of a duration and number of people risk determination.

Risk class (from table above) / ***Proximity***

	Low	Medium	High
High	Risk Priority 2	Risk Priority 1	Risk Priority 1
Medium	Risk Priority 3	Risk Priority 2	Risk Priority 1
Low	Risk Priority 3	Risk Priority 3	Risk Priority 2

Figure 11.2 An example of a risk class and proximity risk determination.

Table 11.1 An Example of Viables' Number of Samples Taken and Current Rate of Occurrences

Process	Number of Sampling Events	Number of Locations
Upstream process	375	23
Downstream process	137	15
Filling process	49	4
Total number of samples		2628
Percent of occurrences		0.00
Percent with even one occurrence		0.04

Table 11.1 shows an example summary for viables' results for EM for a hypothetical vaccine manufacturer for 2 years. A number of microbial samples taken during this 2-year period which included multiple manufacturing campaigns were 2628. No occurrences of out-of-specification (OOS) were found. Therefore the rate of possible viable OOS occurrences is <0.04% which is well within recommendations of United States Pharmacopoeia (USP) <1116> Aseptic Processing Environments (for hypothetical ISO6, for instance, the rate should below 3%). Therefore there were no risks associated with viable organisms were currently observed. Subsequently we recommend reviewing Personnel Monitoring and Nonviables' data were to continue our holistic risk-based approach to EM data evaluation.

Personnel can have a substantial impact on the quality of the environment in which the sterile product is processed. Therefore a vigilant and responsive personnel monitoring program must be continually reviewed and trended to predict any possible shifts or changes.

Typically this monitoring is accomplished by obtaining surface samples of each aseptic processing operator's gown locations at least daily, or in association with each batch. This is accompanied by an appropriate sampling frequency for strategically selected locations of the gown, along with right and left glove fingers (hood, chest, right and left forearm and right and left upper thigh). In addition, quality system must be established with a comprehensive monitoring program for operators involved in operations which are especially labor intensive, that is, those requiring repeated or complex aseptic manipulations. An example of Personnel EM sampling for a hypothetical vaccine company for a period of 2 years where analyzed for each of the processes which included at least glove (fingers) and presented in the following statistical and microbiological assessment. An example rate of occurrence is determined to be 0.565% which is well within recommendations of USP <1116> Microbiological Controls and Monitoring of Aseptic Processing Environments for ISO 5 (<1%), for ISO 6 (<3%). It could be theorized, based on data, that correlation may exist between frequency of sampling and number of OOS occurrences. In author experience, at least empirically, frequent sampling will result in a more frequent failure until actual variability level is identified. When analyzed hypothetical data seem to point to human origin of OOS occurrences rather than equipment, environment, or specific process, based on statistical analysis and recovered organisms.

Tables 11.2 and 11.3 and Figs. 11.3–11.7 summarize results of analyses.

Additionally, typical recovered organisms in this example belonged to two species *Bacillus* and *Staphylococcus*, which strongly suggest operator origins of these OOSs. Contamination during sampling or testing also cannot be ruled out.

Table 11.2 An Example of Personnel (Glove) Microbial Monitoring Summary

Measuring Matric Description	Metric
Number of processes sampled	10
Number of operators sampled	45
Number of samples taken	1948
Number of bioburden occurrences	11
% Occurrence	0.565

Table 11.3 An Example Sampling Summary—Operators Rate of Occurrence

Operators	% Occurrences	Operators	% Occurrences	Operators	% Occurrences
A	1	P	0	AE	0
B	0	Q	0	AF	25
C	0	R	2	AG	0
D	0	S	0	AH	0
E	0	T	0	AI	0
F	0	U	0	AJ	0
G	1	V	0	AK	0
H	0	W	0	AL	1
I	0	X	0	AM	3
J	0	Y	0	AN	0
K	0	Z	0	AO	0
L	0	AA	0	AP	0
M	0	AB	0	AQ	0
N	0	AC	2	AR	0
O	4	AD	0	AS	1

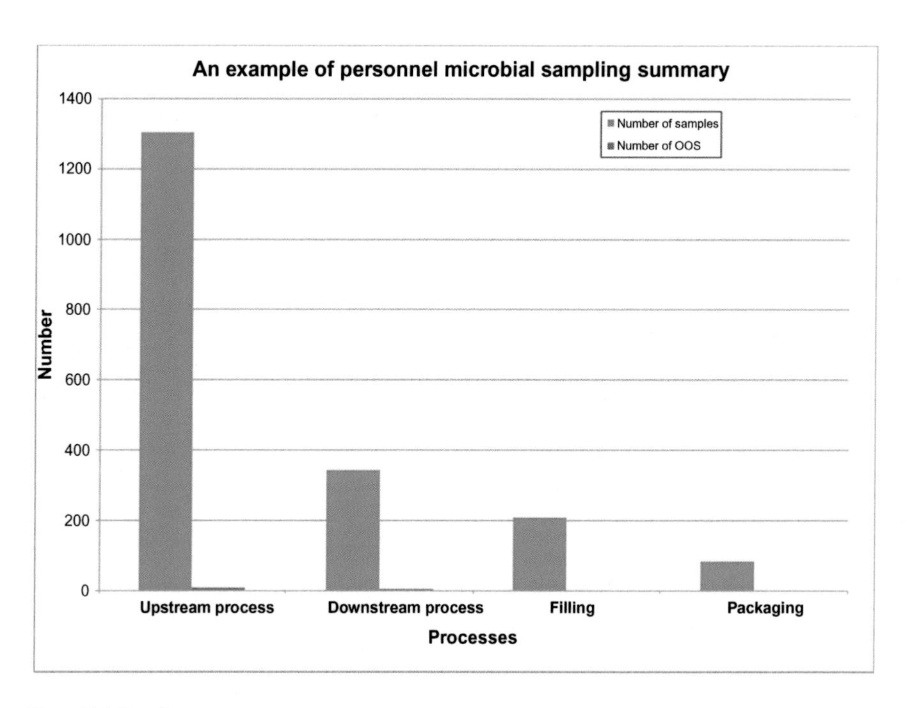

Figure 11.3 Sampling summary—processes.

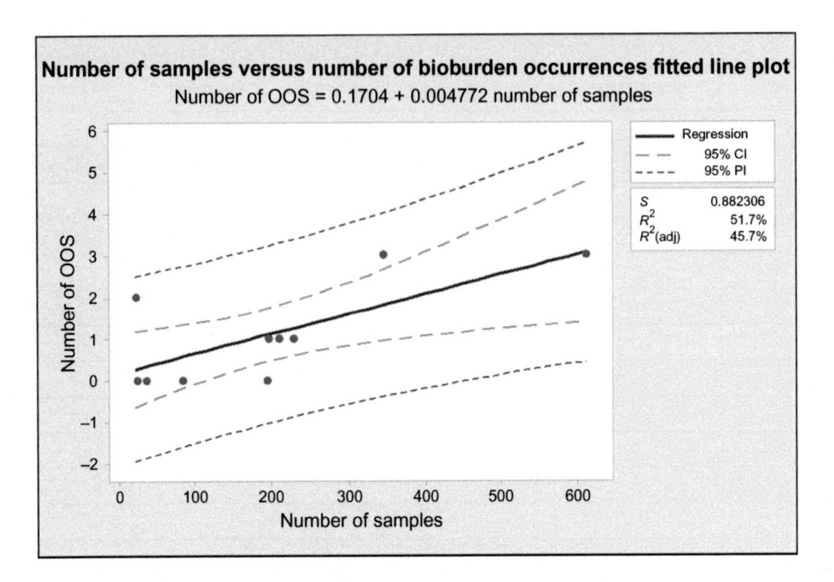

Figure 11.4 An example sampling summary—sampling frequency versus occurrences.*
**A weak correlation ($R^2 = 51.7\%$) is observed between number of samples taken per process and number of bioburden occurrences. This may suggest that the more samples are taken the higher the probability of bioburden occurrences.*

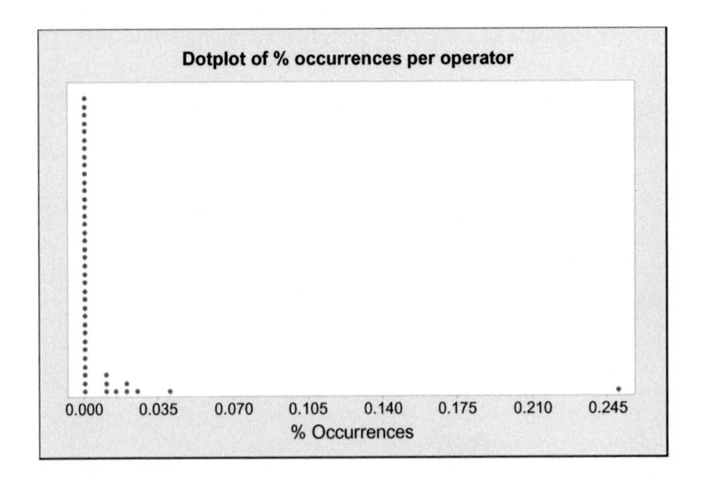

Figure 11.5 An example of sampling summary: % occurrences per operator.

Above is an example of extensive analyses which allows us to comprehensively evaluate microbial load in our process area which in turn supports aseptic nature of a parenteral manufacturing process as well as sterility of our product.

Nested ANOVA: Rate of OOS Occurrence versus Process, Operator

Analysis of Variance for Rate of OOS Occurrence

Source	DF	SS	MS
Process	9	0.0293	0.0033
Operator	81	0.0454	0.0006
Total	90	0.0747	

Variance Components

Source	Var Comp.	% of Total	StDev
Process	0.000	37.17	0.018
Operator	0.001	62.83	0.024
Total	0.001		0.030

Expected Mean Squares

1	Process	1.00(2) + 8.11(1)
2	Operator	1.00(2)

Figure 11.6 An example of a fully nested ANOVA analyzing times OOS occurred versus process and operator.*
**An example multivariate analysis performed using Fully Nested ANOVA analyzing Rate of Occurrence versus Process and Operator. This analysis also suggested that occurrences are more operator related rather than process related as Process contributed 37.17% to Variance, while Operator contributed 62.83%.*

Nested ANOVA: Times OOS versus Process, Operator

Analysis of Variance for Times OOS

Source	DF	SS	MS
Process	9	1.7624	0.1958
Operator	81	11.9079	0.1470
Total	90	13.6703	

Variance Components

Source	Var Comp.	% of Total	StDev
Process	0.006	**3.93**	0.078
Operator	0.147	**96.07**	0.383
Total	0.153		0.391

Expected Mean Squares

1	Process	1.00(2) + 8.11(1)
2	Operator	1.00(2)

Figure 11.7 Rate of occurrence versus process, operator.*
**An example multivariate analysis performed using Fully Nested ANOVA analyzing Times OOS occurred versus Process and Operator. This analysis suggested that occurrences are operator related rather than process related as Process contributed 3.93% to Variance, while Operator contributed 96.07%.*

As stated above this example includes also evaluation of results of must variable size nonviable particles—0.5 μm. The results for a hypothetical vaccine manufacturer's EM 2-year program from each of the Processes were analyzed and tabulated. The Statistical Capability Analyses and Run Charts were used to analyze and compare data. This evaluation is not much different as one discussed in Process Validation chapters of this book.

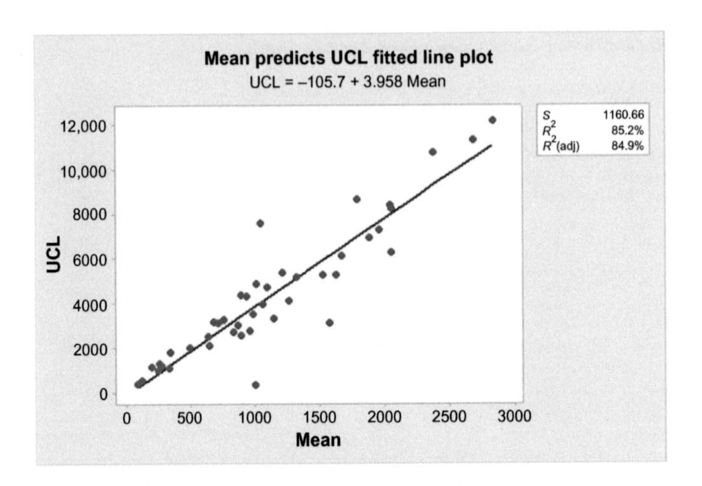

Figure 11.8 Nonviable results mean of the data sets versus upper control limit comparison.

One interesting evaluation from this analysis should be noted. Namely a mean of nonviable results data sets versus upper control limit comparison was performed and is shown in Fig. 11.8.

Fig. 11.8 upon completion of the analyses for nonviable 0.5 μm particles monitoring results for the hypothetical parenteral product's processes, found to be predictable as means of the data sets values consistently correlated with Upper Control Limits (UCL) (regression with $R^2 > 80\%$). All results for 0.5 μm particles were well within specifications with rare events that came to proximity to ISO Class 6 Specification during Upstream Processes, which is expected and by far lower risk than Downstream or Filling Processes.

In addition, the firms must review their EM Program for compliance with ISO 14644-01 (Second edition December 15, 2015) sampling location per this standards Table A. We are recommending using the results of the above analyses to confirm or update existing locations and to identify new locations, as required.

In conclusion we want to once again mention the following several concepts:

- Clean rooms must be designed to commensurate the risk with regard to the process, especial consideration should be given to airflow and filtration.

- Consider typical major sources of contamination which are personnel, materials, equipment and incoming air (in that order).
- Consider analysis of isolated organisms to understand sources of their origin so that strategy for their elimination and prevention could be effectively implemented.
- Consider an importance of a statistical analysis in evaluation of science and risk-based EM program, as sampling prescribed in ISO 14644 is statistically based.
- Design the program to be agile based on shifts in the process.
- Consider the incubation time for viables' samples on the type of a media used and type of organisms planning to recover.[3]
- Consider the growth promotion studies such as protocol for USP and House recovered organisms to qualify viables' sampling.[4]
- Consider instituting a nonviable particle monitoring and trending inside Biological Safety Cabinets during routine manufacturing operations by establishing a data analyses base line and performing statistical comparison of data attained.
- Consider establishing an appropriate sampling frequency for all strategically selected locations of the gown, along with right and left glove fingers (hood, chest, right and left forearm and right and left upper thigh) will be established based on outcomes of the personnel monitoring data review.
- Consider evaluation and establishment of additional and periodic training at the start of work of new employees.

Finally, a well-established EM program assures deep understanding of variability due to this factor and thus provides a great source of knowledge to Process Validation program for parenteral products manufacturing.

References

Moldenhauer, J., Madsen, R., Contamination Control in Healthcare Product Manufacturing, vol. 3.

EudraLex, 2008. The Rules Governing Medicinal Products in the European Union, vol. 4, EU Guidelines to Good Manufacturing Practice Medicinal Products for Human and Veterinary Use, Annex 1, Manufacture of Sterile Medicinal Products (corrected version), Brussels, 25 November (rev.).

[3] For instance, TSA (TAC) media requires 48–96 hours at 30°C–35°C, while SDA (fungi) requires 5–7 days at 20°C–25°C. The type of agar and the incubation conditions will also affect the types of microorganisms recovered.

[4] Since EM is a Risk and a Science Based program's data should be evaluated on periodic basis, the microorganisms' isolates should be trended to see if there are more studies required.

FDA, 2004. Guidance for Industry, Sterile Drug Products Produced by Aseptic Processing—Current Good Manufacturing Practice, U.S. Department of Health and Human Services, Food and Drug Administration, Center for Drug Evaluation and Research (CDER), Center for Biologics Evaluation and Research (CBER), Office of Regulatory Affairs (ORA), September.

Imai, M., 1997. Gemba Kaizen: A Commonsense Low-Cost Approach to Management. McGraw-Hill Professional, New York, p. 13. ISBN: 978-0-07-031446-7.

International Standard Organization, Aseptic Processing of Health Care Products, Part 1, General Requirements, International Standard ISO 13408-1.

International Standard Organization, 2015. Cleanrooms and Associated Controlled Environments, Part 1, Classification of Air Cleanliness, International Standard ISO 14644-1.

Ishikawa, K., 1976. Guide to Quality Control. Asian Productivity Organization, ISBN: 92-833-1036-5.

PDA DHI Technical Books, Chapter 11: Hal Baseman and Mike Long-Intervention Risk Assessment Model (IREM).

Stabler, M., Famili, P., 2015. Risk Based Environmental Monitoring (EM) and EM Data Management and Trending an Industry Roundtable Discussion, March. PDA New England Chapter Meeting Presentation.

Further reading

Baseman, H.S., Stabler, M., Henkels, W., Long, M., 2016a. A Line of Sight Approach for Assessing Aseptic Processing Risk: Part I, PDA Letter. May 31.

Baseman, H.S., Stabler, M., Henkels, W., Long, M., 2016b. A Line of Sight Approach for Assessing Aseptic Processing Risk: Part II PDA Letter. July 07.

Baseman, H.S., Stabler, M., Henkels, W., Long, M., 2016c. A Line of Sight Approach for Assessing Aseptic Processing Risk: Part III PDA Letter. September 27.

International Standard Organization, Cleanrooms and Associated Controlled Environments, Part 2, Specifications for Testing and Monitoring to Prove Continued Compliance with ISO 14644-1.

International Standard Organization, Cleanrooms and Associated Controlled Environments, Part 14, Design, Construction and Start Up of Clean Room Facilities, International Standard ISO 14644-4.

CHAPTER 12

Isolators

Chris Smalley
ValSource Inc., Downingtown, PA, United States

Isolators are currently considered the best practice for aseptic and low particulate operations. For "aging facilities," isolators can be part of the solution to upgrade the facility. Isolators represent an opportunity to take a Class A/Class 100 space and improve the level of assurance that an aseptic environment is being maintained.

Isolators have a robust record of use in testing laboratories, and that experience points up the challenges that needs to be borne in mind when constructing a validation package. Those challenges include:

- *Access*—the design of the isolator can limit operations within the unit as well as limit access.
- *Throughput*—extended isolator preparation and end of campaign activities can increase production availability.
- *VHP Decontamination*

The design of an isolator is based on the encapsulation of air using stainless steel, plastic, and glass, with provisions for handling materials by gloves via glove sleeves or half-suit(s). A microbial retentive filtration system using High Efficiency Particulate Air (HEPA) provides clean air and is able to maintain a positive pressure as required.

The operator gains access to the material inside the isolator by gloves and glove sleeves, thus the operator remains physically within, but biologically separated from the isolator's internal environment. Isolators do not allow the ingress of airborne contamination (as limited by the HEPA filter efficiency) from the surrounding environment. Introduction of personnel borne contamination into the isolator is precluded. The isolator can be opened to the surrounding room by means of hatchback window(s) that allows entry/egress of materials into or out of the isolator, when necessary.

Principles of Parenteral Solution Validation. DOI: https://doi.org/10.1016/B978-0-12-809412-9.00012-5

The isolator can be processed by Vapor Phase Hydrogen Peroxide (VHP) using a reproducible cycle that biodecontaminates the exposed, internal surfaces of the isolator and the exposed, external surfaces of materials and equipment placed within the isolator.

The VHP Generator is a standalone unit that forms a closed-loop with an isolator and uses HEPA-filtered air as a carrier to deliver hydrogen peroxide gas to the exposed surfaces inside the sealed isolator. The oxidative nature of hydrogen peroxide gas forms the mechanism of lethality.

Hydrogen peroxide gas is continuously injected for the required exposure time to assure proper decontamination. Once the hydrogen peroxide gas leaves the enclosure, it is catalytically converted into water vapor and oxygen. The water vapor is stored in the Generator's drying system.

Equipment/components identified for use within the isolator during filling operations decontamination using the isolator will be washed and prepared following operational procedures. The items will then be loaded into the isolator with the location and orientation of each item defined. With all items loaded per approved load configuration, the isolator door will be closed, and the cycle started. The decontamination process consists of four phases: Dehumidification, Conditioning, Decontamination, and Aeration.

Verify that all IQ/OQ protocols and studies associated with the {System} and critical supporting utilities have been completed. Verification should be documented as a precursor to study execution activities and captured in the execution phase of this study protocol. Any open exceptions identified under the IQ/OQ should be reviewed to ensure these do not impact the expected performance of the system's decontamination cycle. Any deviations documented during the validation activities listed as prerequisites must have been resolved.

Understanding the basics of the design, the challenges mentioned earlier deserve a detailed discussion. *Access*—access to the Isolator, once closed and the VHP cycle has been completed, is by use of the gloves and, where installed, pass-through. The Isolator must be demonstrated to be leakproof for execution of the VHP cycle, and leakproof during operation unless designed for use with a "mouse hole" for product to leave the isolator. When designed with a mouse hole, the design must include a leakproof door for the VHP cycle, and the amount of positive pressure and air loss through the mouse hole controlled to maintain Isolator integrity. These attributes will need to be tested during Validation. Procedures need to be in place and personnel trained to assure that no inappropriate access occurs once the VHP cycle has executed.

Throughput—Once operations are complete, the Isolator will need to be opened, materials associated with the previous lot removed and cleaning performed. Cleaning will need to accomplish three purposes, removal of all contamination introduced by the previous lot, elimination as much as possible of microbial contamination, and elimination as much as possible of particulates. For the latter purpose, it is important that the cleaning process, through the use of cleaning wipes and other materials, does not itself become a source of particulates into the Isolator. Once cleaning is completed, change parts and components that can be subject to the VHP cycle should be installed and the Isolator closed. There are no shortcuts in Isolator cleaning and set-up, nor in the VHP cycle that follows. Cleaning attributes will need to be tested during Validation. Those that are manual will need procedures and training in place.

VHP Decontamination—There is no reliable model for predicting lethality from physical measurements for VHP as is possible for moist heat sterilization methods, such as Fo. An additional challenge using VHP is residual peroxide that may impact products susceptible to oxidation. VHP is a surface decontaminant. The quality of Biological Indicators (BIs), which are a direct measurement of the cycle efficiency, may affect the implied resistance to the VHP cycle. One BI failure in a Validation Study has the potential to halt production.

Focusing on the BI, debris among the spores, or clumping of the spores can result in some spores being "protected" from VHP exposure. The debris might be media. The possibility of unexpected growth by a BI during studies which are not due to process lethality failures, but rather to the "protected" spores, gives rise to the term, "rogue BI." Replicate BI strategy significantly reduces the potential of a false positive through the placement of three BIs at every test location.

BIs need to be the subject of a supplier evaluation. The criteria for this evaluation includes: (1) compendial testing for *Geobacillus stearothermophilus* consistent with USP <1035> with a count of $1-5 \times 10(6)$ spores, (2) D-value of 0.8–3.0 minutes, (3) total kill time test which is 20 times the manufacturer's D-value to assure no outliers, and when investigation is necessary, (4) SEM—physical examination of sample BIs by Scanning Electron Micrographs to assure no clumping or atypical BIs.

Once certain that the BIs are capable of providing the quality information needed for the Validation, the study can commence. The Isolator manufacturer will provide an initial series of settings for the VHP cycle based on dehumidification of the Isolator, Conditioning of the Isolator, Decontamination where the VHP is exposed, and finally Aeration to reduce the level of VHP to acceptable levels. If these cycles have not been exercised in OQ, they will need to be exercised prior to the introduction of BIs to check for cycle operation. When ready for testing, the following criteria and documentation of chemical indicator, Draeger Tube (or other physical test method for VHP exposure) and BIs are prepared, and the testing conducted (refer to Tables 12.1–12.4).

Table 12.1 Validation Criteria

BI Recovery Evaluation (Total Kill)	BIs at all locations must be recovered for the study to be valid for the Total Kill approach.
BI Recovery Evaluation (Replicate BI)	If one BI out of the three BIs in one location is not successfully recovered, this by itself should not invalidate the run but instead the study conclusion should be based on the assumption that the BI is positive.
BI Positive Control	Shows growth
BI Negative Control	Shows no growth
Draeger VHP concentration	

Table 12.2 Example Chemical Indicator Locations

CI Number	Location	Rationale for Location
1	Bottom right corner of isolator	Lowest temperature during temperature profiling
2	Entry to isolator	Highest temperature during temperature profiling
3	Within isolator glove 4	Lower smoke dispersal during airflow visualization
4	Within Isolator glove 8	Lower smoke dispersal during airflow visualization
5	Top right corner	
6	At Rapid Transfer Port	

Table 12.3 Example Biological Indicator Locations

TC Number	Location	Rationale for Location
1	Top left corner	General geometric challenge
2	Top right corner	General geometric challenge
3	Bottom left corner	General geometric challenge
4	Bottom right corner	General geometric challenge
5	Upper Plane	General geometric challenge

Table 12.4 Example Draeger Tube Locations

Location Number	Location	Rationale for Location
1	Infeed—Operator Side	The six sample locations are near the inlet return of the recirculating air and represent a complete and even distribution across the isolator
2	Middle—Operator Side	
3	Outfeed—Operator Side	
4	Infeed—Back Side	
5	Middle—Back Side	
6	Outfeed—Back Side	

Validation of Isolator, especially the VHP cycles, is considered in the same manner as steam sterilizers—it should be repeated on no less than an annual basis. For that reason, rather than constructing a protocol for testing, consideration should be given to writing a procedure for the consistent, periodic testing, and the reporting of that testing.

Further Reading

EU Guidelines, 2008. to Good Manufacturing Practice Medicinal Products for Human and Veterinary Use Annex 1, Manufacture of Sterile Medicinal Products (corrected version) EudraLex The Rules Governing Medicinal Products in the European Union, Volume 4, Brussels, 25 November (rev.).

FDA, 2004. Guidance for Industry, Sterile Drug Products Produced by Aseptic Processing — Current Good Manufacturing Practice, U.S. Department of Health and Human Services Food and Drug Administration, Center for Drug Evaluation and Research (CDER), Center for Biologics Evaluation and Research (CBER), Office of Regulatory Affairs (ORA), September, Pharmaceutical CGMPs.

CHAPTER 13

Post Aseptic Fill Sterilization and Lethal Treatment

Michael J. Sadowski
Lead Scientist, Sterility Assurance, Baxter Healthcare Corporation, IL, United States

Introduction

Due to inherently lower risks of nonsterility, terminal sterilization processes are universally considered to be superior to aseptic processes and thus preferred by regulators. However some parental drug product formulations are not stable and may degrade due to processing by terminal sterilization processes that utilize exposure to moist heat. Accordingly these products necessitate the ongoing need for aseptic processing.

With aseptic processing, there is a very low probability that any filled unit might contain a microorganism due to the robust process design and associated environmental microbial controls. Any microorganisms present are low in number and most likely from human origin (i.e., primarily Gram-positive cocci) and are expected to demonstrate relatively low resistance levels to moist heat sterilization processes.

To further reduce the risk of nonsterility with an aseptic product that can tolerate some of level of heat, a post aseptic fill lethal treatment process could be employed. For this situation, the product is aseptically filled and sealed in containers followed by exposure to a validated microorganism inactivation process that delivers a low level of heat history as the terminal step of the overall manufacturing process. Based on the heat sensitivity of the product, the low heat history treatment step can be performed at conventional moist heat sterilization temperatures or at lower temperatures.

It is important to note that the application of a post aseptic fill lethal treatment process is not to be considered mandatory. Based on the documented, successful, safe application of aseptic processing for many years, there is a

Principles of Parenteral Solution Validation. DOI: https://doi.org/10.1016/B978-0-12-809412-9.00013-7

lack of scientific and risk-based evidence to support the absolute need for the application of terminal sterilization or other lethal treatment processes for product produced via well designed, properly controlled and operated aseptic processes. Accordingly, aseptic manufacture in these cases can provide products of suitable quality and there is no scientific justification to support the expectation that products produced through aseptic manufacture would necessarily require the addition of some moderated 'terminal sterilization' or other lethal treatment conditions.[1]

The implementation of a post aseptic processing sterilization or lethal treatment process step yields a reduced risk for nonsterility. Therefore optimization of the sterility assurance program should be one of the benefits derived from this enhanced focus on proactive measures to support product sterility. This chapter will discuss lethal treatment approaches, considerations, and potential benefits that can be derived from applying a terminal microbial inactivation process on product that has been previously filled using an aseptic process.

Sterilization and Lethal Treatment

The term "sterilization" is a well-known term that is very precisely defined in global regulatory standards and guidance documents while the term "lethal treatment" is a much broader term that is not as precisely defined nor widely understood. Accordingly the author prefers to recognize the terms "post aseptic fill sterilization" and "post aseptic fill lethal treatment" in a distinctive and exclusive fashion much in the way that traditional sterilization is not commonly referred to as lethal treatment.

Post Aseptic Fill Sterilization

The following definition and clarifying notation for the term sterilization can be found in ISO11139[2] (2018):

> *Sterilization*: Validated process used to render product free from viable microorganisms.
> *Note*: In a sterilization process, the nature of microbial inactivation is described as exponential and thus the survival of a microorganism on an individual item can be expressed in terms of probability.

[1] Adapted from PDA Comments Submitted to EMA on "Guideline on the sterilization of the medicinal product, active substance, excipient and primary container," October 2016.

[2] ISO11139 (2018). Sterilization of healthcare products–Vocabulary–Terms used in sterilization and related equipment and process standards.

While this probability can be reduced to a very low number, it can never be reduced to zero.

Aseptic processes provide a high level of bioburden control based on the robust and successful design, development and qualification (e.g., Aseptic Process Simulation, Environmental Monitoring Performance Qualification, etc.) of the process. The resulting low product bioburden levels support the use of the Product Specific Design Approach and an associated reduced physical lethality (F_0) for moist heat sterilization processes for any products that can tolerate a relatively low heat history at recognized sterilization temperatures. Since the presence of product bioburden in this situation should be at a low probability, reduced assumptions for the product bioburden population level and resistance level may be considered in the determination of the Probability of a Nonsterile Unite (PNSU) for these products. Additionally, the use of the Product Specific Design Approach could provide for a PNSU of $\leq 10^{-6}$ which represents a significant reduction of the risk of nonsterility when compared to traditional aseptic processing. The use of the Semilog Survivor Curve Equation adapted from PDA Technical Report No. 1 Eq. (13.1)[3] can be used to calculate the resulting PNSU for product with exposure to a low heat history sterilization including the product bioburden assumptions listed below:

$$\text{Log } N_F = -F_0/D_{121^\circ\text{C}} + \text{Log } N_0 \tag{13.1}$$

where, N_F is the number of product bioburden microorganisms surviving after exposure to a specific F_0. N_F is also the PNSU. The target for a terminal moist heat sterilization process is a PNSU $\leq 10^{-6}$. F_0 is the equivalent lethality of a cycle calculated as minutes at a reference temperature of 121°C, using a defined temperature coefficient or z-value of 10°C. F_0 should be calculated from heat penetration temperature readings in the slowest to heat location in the product. Assume an F_0 value of 2.4 minutes for this example. $D_{121^\circ\text{C}}$ is the $D_{121^\circ\text{C}}$ value—thermal resistance value, in minutes, of a product bioburden microorganism at 121°C. Assume a product bioburden resistance of 0.4 minutes for this example. N_0 is the number of product bioburden microorganisms prior to exposure. Assume 1 heat resistant spore for product bioburden for this example.

[3] PDA Technical Report No. 1 (2007). Validation of Moist Heat Sterilization Processes: Cycle Design, Development, Qualification and Ongoing Control.

Then,

$$\text{Log } N_F = -F_0/D_{121°C} + \text{Log } N_0$$

$$\text{Log } N_F = -2.4/0.4 \text{ minutes} + \text{Log }(1) = -6 + 0$$

$$\text{Log } N_F = -6$$

$$N_F = 10^{-6}$$

Note: An additional safety factor can be added by increasing the F_0 value to 3 minutes or more to further reduce (enhance) the resulting PNSU for the post aseptic fill moist heat sterilization process.

In order to support the achievement of a corresponding moist heat sterilization process biological lethality (F_{BIO}), biological indicators (BI's) with $D_{121°C}$ values in the range of 0.2–0.6 minutes (e.g., *Bacillus subtilis* 5230) are available from various BI manufacturers.

F_{BIO} is calculated as follows:

$$F_{BIO} = D_{121°C}(\text{SLR}) \tag{13.2}$$

where, F_{BIO} is the biological lethality for the sterilization process; $D_{121°C}$ is the $D_{121°C}$ value of the BI; SLR is the Spore Log Reduction (SLR) achieved for the BI after exposure to the sterilization process.

$$\text{SLR} = \text{Log}(N_0) - \text{Log}(N_F) \tag{13.3}$$

where, N_0 is the starting spore population of the BI prior to exposure to the sterilization process and N_F is the surviving spore population of the BI after exposure to the sterilization process. An SLR value of 6.0 will be used for this example with the assumption of complete inactivation of a BI with a starting population of 10^6 spores per unit. In this case, an N_F of 1 must be assumed since it is not mathematically valid to calculate the logarithm of 0.

Then $\text{SLR} = \text{Log }(10^6) - \text{Log }(1) = 6.0 - 0$

$$\text{SLR} = 6.0$$

Then,

$F_{BIO} = D_{121°C}$ value (SLR)
$F_{BIO} = 0.4$ minutes (6.0)

$F_{BIO} = 2.4$ minutes which can be included in place of F_0 in Eq. 13.1 above to also support a PNSU $\leq 10^{-6}$.

In addition to or as an alternate to moist heat sterilization, other modalities of terminal sterilization such as radiation (gamma or e-beam) may also be used to enhance the PNSU/product sterility after completion of aseptic processing.

Post Aseptic Fill Lethal Treatment

Post aseptic lethal treatment can be defined as the application of a terminal treatment process designed to inactivate microorganisms present after completion of an aseptic process. As stated above and due to the extensive environmental controls and the prevalence of microorganisms from human origin that are vegetative microorganisms and not spore-formers, the number and resistance of any microorganisms present in product from aseptic processing is expected to be very low.

In order to inactivate vegetative organisms to obtain a count that is exclusively from spores in the Spore Count Test, a heat shock at 60°C–80°C for 15–30 minutes is often employed on the samples prior to plating and incubation. Therefore these temperatures and exposure ranges can also be considered for use with post aseptic fill lethal treatment for those products that are not negatively affected by exposure to these levels of heat history. Based on this author's experience, 80°C for 15 minutes seems to be a widely used heat history for sanitization which could be suitably adapted for post aseptic fill lethal treatment. However the temperature set point should be based on the heat sensitivity of the product while the overall target heat history should be based on the resistance of the microflora at the aseptic processing facility. This could lead to potential higher or lower temperature set points with shorter or longer exposure times, respectively.

The simplest heat processing option for the post aseptic fill lethal treatment would be heat exposure in an appropriately sized water bath in a batch type of process although it would also be possible also utilize a continuous process with appropriate equipment. The process heat history could be set and demonstrated by monitoring time at temperature (i.e., 15 minutes at 80°C) or with integration of time and temperature similar to the approach with F_0 but with an alternate reference temperature (T_{ref}) corresponding to the target exposure temperature. For temperature monitoring, the temperature could be monitored in the slowest to heat location of product for each batch or the time needed at the minimum exposure temperature could be developed and qualified.

Although this section discussed the application of moist heat with a water bath, other sterilization technologies (e.g., radiation) could be similarly utilized based upon the sensitivity of the product to these technologies.

Parametric Release

As stated previously, the implementation of proactive measures such as post aseptic fill sterilization and lethal treatment should result in an optimized sterility assurance program. The application of parametric release aligns well with the implementation of proactive and validated microbiological inactivation processes which far exceed the risk mitigation potential of highly intensive product testing strategies such as the finished product sterility test. It is well known that the product sterility test is statistically flawed with very limited capability to support that a batch of moist heat terminally sterility product is sterile. It is important to note that these same limitations also exist with the finished product sterility test used with aseptically filled products.

PDA Technical Report No. 30[4] defines parametric release as: "A sterility release program that is based on effective control, monitoring and documentation of a validated sterile-product manufacturing process where sterility release is based on demonstrated achievement of critical operational parameters in lieu of end-product sterility testing." FDA has stated that a firm will be approved for parametric release based on how well the firm has mitigated all risks related to product sterility.[5] Accordingly, it could be possible in the future for a parenteral product that utilizes a validated post aseptic fill lethal treatment process to reduce the risk of nonsterility to be a be a potential candidate for parametric release.

Post aseptic fill lethal treatment has a critical role to play in the pursuit of Aseptic Parametric Release. As stated above, it is well

[4] PDA Technical Report No. 30 (2012). Parametric Release of Pharmaceutical and Medical Device Products Terminally Sterilized By Moist Heat.

[5] FDA Submission Guidance Document (2010). Submission of Documentation in Applications or Parametric Release Of Human and Veterinary Drug Products Terminally Sterilized with Moist Heat Processes.

established that the sterility test is limited in its ability to reliably and accurately confirm that aseptic products are sterile. In consideration of the proposition of parametric release with an aseptic product exposed to a post aseptic fill lethal treatment, the role of the sterility test could potentially be eliminated and replaced by the adoption of relevant traditional parametric release practices (i.e., comprehensive risk assessment to identify and mitigate all risks to sterility, etc.) including the critical parameters of the lethal treatment process. In comparison to conventional aseptic processing with sterile product release supported by a finished product sterility test, it is clear that an aseptic product subjected to a validated post aseptic fill lethal treatment and administered by a parametric release sterility assurance program represents a superior proposition.

Summary

The implementation of a post aseptic fill sterilization or lethal treatment process drives an enhanced product PNSU and a superior sterility assurance program when compared to conventional aseptically filled sterile products. With the application of post aseptic fill sterilization or lethal treatment as a terminal step to further reduce process and product nonsterility risk, other potential benefits that could be considered including reduction in environmental monitoring, the need for fewer media-fills and even the possibility for the adoption of parametric release.

CONCLUSION

We would like to thank the reader for choosing our book. In this book we tried to review current regulatory and industry guidance for establishing Process Validation programs for Parenteral products. As we stated in the Introduction section of this book "in addition to overview of a lifecycle of the parenteral product processes which include discussions about process design, process qualification and process maintenance will also touch upon subjects that help specifically for aseptic process manufacturing such as use of statistics, aseptic process simulations, post aseptic sterilization and others." When we embarked on this writing journey our goal was to present parenteral drug manufacturers with a volume that helps them implementing holistic and comprehensive process validation programs. We also wanted to design a "one stop shop" on the subject.

Prior to start of this book we surveyed industry experts what they would like to see and what would be valuable to them in this book.

Here are some of the topics they wanted us to cover:

- The American perspective on the interpretation of the FDA's Guidance for Industry on Process Validation.
- Different approaches to problem solving in this area.
- Regulatory considerations.
- Risk management exercises.
- Preparation and execution of properly designed studies.
- Assistance in scale-up and technology transfer activities.
- Focus on examples and case studies.

We are happy to state that we covered all of these topics. We started with a history of the subject of Process Validation and it current state. Than we proceeded to subjects that are perquisites and are supplemental subjects required to be established and in a state of control to be successful when implementing Process Validation program. These included:

- Quality Systems;

- Aseptic Process Design and Planning;
- Quality Risk Management;
- Contamination Control and Cleaning Validation;
- Statistics.

Then actual three-stage approach to process validation was covered including:

Stage 1—Process Design:

- Master planning, organization, and schedule planning;
- Risk and impact assessment.
- Process and systems design;
- Perform risk assessment (Identification of Critical Quality Attributes and Critical Process Parameters);
- Design of Experiments;
- Developing control strategies and determine process design;
- Scale-up;
- Reporting and tech transfer strategy.

Stage 2—Process Performance Qualification:

- Clean room design and operation, User Requirements and Equipment Qualification.
- Determine Test Plan and Acceptance Criteria for PPQ, Sample Size, and Number of Batches.
- Protocol development.
- Analysis and interpretation of results.

Stage 3—Continued Process Verification

- Determining when CPV starts;
- Legacy systems vs. new systems;
- CPV strategy and enhanced sampling;
- Maintenance of Validation and Change Control, and Periodic Assessment;
- Develop strategies for CPV.

In addition, we covered Special Topics, such as Preuse/poststerilization Integrity Testing of Sterilizing Grade Filter, Risk Based Environmental Monitoring, Aseptic Processing Simulation, Isolators and Post Aseptic Lethal Treatments Post Aseptic Fill Sterilization and Lethal Treatment.

Our hope that reader finds this volume useful in their implementation of Process Validation programs for Parenteral pharmaceutical products.

Igor Gorsky
Senior Consultant, ConcordiaValsource LLC.,
Downingtown, PA, United States

INDEX

Note: Page numbers followed by "*f*" and "*t*" refer to figures and tables, respectively.

A

ABCtiter, 162, 166
Aborted and invalid media fills, 55
Acceptable Daily Exposure (ADE), 83–84, 83*t*
Acceptable Daily Intake (ADI), 83*t*
Acceptance criteria, 53, 213
Active Pharmaceutical Ingredients (API), 85, 221
  Subject to Pre-Market Approval, 5
Actual process capability (Ppk), 228
Aeration, 256
Affinity diagrams, 75–76
Agitation calculation procedure, scale, 160*t*
Airflow patterns in isolators, 191–192
Anaerobic processes, 57
Annual or periodic production quantities, 21
Annual product review (APR), 229–230
API. *See* Active Pharmaceutical Ingredients (API)
Aseptic practices, 28–30
Aseptic process simulations
  aborted and invalid media fills, 55
  acceptance criteria, 53
  anaerobic processes, 57
  duration, 43–46
  failure investigation, 54–55
  filled unit accountability, 54
  fill volume, 43
  growth promotion studies, 51–52
  incubation, 50–51
  intervention evaluation and risk assessment methods, 47–50
  interventions, 46–47
  lyophilized product filling, 56–57
  media fills run number, 39–41
  microbial contamination, 35–38
  ointment filling, 56
  performance schedule and frequency, 39
  postincubation inspection, 53
  powder filling, 56
  preincubation inspection and rejection, 52–53
  process steps, inclusion, 41–42
  study, 38–39
  validation, 34
  "worst-case" parameters, 42–43
Aseptic process validation
  aseptic practices, 28–30
  background, 9–12
  basis of design, 21–22
  capability, 14–15
  control, 14–15
  equipment and facility qualification, 23–25
  first air principles, 30–31
  life cycle approach, 16–17
  line of sight (LOS) approach, 17–20
  mapping, 25–27
  performance assurance, 17
  periodic assessment and requalification, 27–28
  reasons for difficulty, 12–14
  requirements, 20–21
  timely user requirement specification, 22–23
  understanding, 15–16
  validation, 14–15
Aseptic technique or practices, poor, 35
Aspergillus brasiliensis, 52
Audits
  frequency using risk matrix, 112
  schedule, 112–113, 112*t*

B

Bacillus subtilis, 195
Basis of design, 21–22
Batch(es)
  data analysis, 134
  number, process validation, 202–205
Batch-to-batch variability, 3–4
Benefit scoring criteria, 103
Biological Indicators (BIs), 257–258
  locations, 259*t*
Biological lethality, 264
Brevundimonas diminuta, 195–196
Bubble Point, 233

C

Candida allicins, 52

CAPA. *See* Corrective and Preventive Actions (CAPAs)
Capability of a process, 213
Change control, 105–106
  corrective and preventive action effectiveness monitoring, 106–107
Checklist, 72
Chemical indicator locations, 258*t*
Chemical sterilization, 193
Cleaning evaluation, 92, 93*f*
Cleaning validation, 89–90, 200
  practitioners, 84
Clean rooms
  and classified areas, 190–192
  environmental deterioration, 44
  personnel, 35
Clean zone layout, 244
% coefficient of variation (%CV), 215
Cohen's rules of thumb for effect size, 204, 204*t*
Commercial manufacturing process, 214
Commissioning, 188
Common-cause variance, 216
Complex risk assessment activity, 68–69
Compliance policy guide (CPG), 5
Component preparation, 192–193
Computer system validation, 200
Concurrent data set, 214
Concurrent release, 214
Conditioning, 256
Confidence intervals (CIs), 90
Confidence level, 214
Confirmation run, 133
Container closure integrity, 200–201
Continued process verification (CPV), 25, 27, 94
  change control and periodic assessment, 225–226
  control rules for, 226–228
  dashboard, 232*f*
  definitions, 213–216
  introduction of, 216–219
  legacy systems *vs.* new systems, 219–221
  maintenance, 225–226
  new product, 220–221
  parenteral products, case study evaluations for, 230–231
  plans, 141
  prerequisites, 219*t*
  process design, 220*f*
  sample process design, 220*f*
  Stage 3a sampling and evaluation, 222*t*
  Stage 3b sampling, evaluation and review periodicity, 223*t*
  strategies for, 228–230
  strategy and enhanced sampling, 221–224
Continuous data, 214
Continuous process improvements, 225–226
Control charting, 123–125
Control limits, 214
Control strategies and determine process design, 156–157
Corrective and Preventive Actions (CAPAs), 40, 98, 102–105
  risk benefit analysis, 102–105
Coverage levels, 203
Coverage probability, 214
Cpk value, 131
CPPs. *See* Critical Process Parameters (CPPs)
CPV. *See* Continued process verification (CPV)
Critical Material Attributes, 146
Critical process parameters (CPPs), 62–63, 86–87, 101–102, 146, 184
Critical quality attributes (CQAs), 17–18, 62–63, 86–87, 101–102, 126, 183–184, 203, 218
  of product, 21, 144
Critical responses, 165*t*
Critical utilities, 189–190
Current Good Manufacturing Practices (CGMPs), 4–5

D

Decontamination, 256
Dehumidification, 256
Depyrogenaiton, 193
Descriptive statistics, 123
Design of experiments (DOE), 86–87, 131–134, 137–138, 151–156
  concepts, 153*f*
  factors summary, 164*t*
  factors *vs.* responses interpretation summary, 167*t*
  mixing study, 154*t*
Design qualification (DQ), 24, 187
Deviation categories, 100, 100*t*
Deviation management system, 99–102, 105–106
  decision tree, 101*f*
  high risk deviation, 101–102
  low risk deviation, 100
  medium risk deviation, 100–101
Diffusive flow, 233
Discrete data, 214

Disposable filters, 197
Documentation, 187
Downstream processing, 144–145
"Drug Product Inspections,", 1–2
Dry heat sterilization, 195

E

EMA, 46
End-to-end assessment, 68–69
Environmental condition-related processes, 12–13
Environmental Monitoring (EM) Program, 13, 243–244
  aseptic processing, 245
  clean zone layout, 244
  concepts, 252–253
  contamination control, 244
  EM-REM, 245–246
  operators rate of occurrence, 249*t*
  out-of-specification (OOS), 247
  people risk determination, 246*f*
  personnel (glove) microbial monitoring, 248*t*
  Personnel Monitoring and Nonviables' data, 247
  rate of occurrence *vs.* process, 251*f*
  risk class and proximity risk determination, 247*f*
  sampling summary—processes, 249*f*, 250*f*
  statistical capability analyses and run charts, 251
  Upper Control Limits (UCL), 252, 252*f*
Equipment cleaning process
  limits establishment, 82–96, 83*t*
  NOEL *vs.* NO[A]EL, 84*t*
  setting cleaning validation limits, risk levels for, 85*t*
Ergonomics, 45
Ethylene oxide, 195
Experimental design, 133
Experimental plan, 133
Expert Working Groups (EWGs), 131–134, 137

F

Factors and factor levels, defining, 133
Factory acceptance testing (FAT), 25, 187
Failure investigation, 54–55
Failure Modes and Effects Analysis (FMEA), 70, 73
Fault tree analysis (FTA), 71–72
FDA, 46
Fermentation
  buffer temperature
    vs. endotoxin, 175*f*
    vs. pentose yield, 169*f*
    vs. residual protein, 172*f*
  duration (post transition phase) *vs.* yield at specific process points, 168*f*
  elution time
    vs. endotoxin, 174*f*
    vs. pentose yield, 168*f*
    vs. residual protein, 171*f*
  endotoxin interaction plots for each of four elutions, 173*f*
  residual protein main effects plots for each of four elutions, 170*f*
  time
    vs. endotoxin, 174*f*
    vs. residual protein, 171*f*
    vs. yield optimization, 169*f*
Filled media, 50–51
Filled unit accountability, 54
Filling equipment and systems, 197–198
Fill volume, 43
Filtration risk assessment, 239*t*
First air principles, 30–31
First in First Out (FIFO) approach, 99
FMEA. *See* Failure Modes and Effects Analysis (FMEA)
FTA. *See* Fault tree analysis (FTA)

G

Gamma irradiation process, 235
Gaussian distribution, 214
General Principles of Process Validation, 178–179
Geobacillus stearothermophilus, 194
GHTF Quality Management Systems, 119
Good manufacturing practices (GMP), 1
Grade C environment, 37
Growth media, 33
Growth promotion studies, 51–52
Guidance for Aseptic Processing, 119

H

HACCP. *See* Hazard Analysis and Critical Control Points (HACCP)
Hazard Analysis and Critical Control Points (HACCP), 70–71
Hazard identification, 75–76
Hazard severity, 148*t*
Health-Based Exposure Limits, 82–83
Health-based exposure limits terms, 83*t*, 86*f*

Heating ventilation and air conditioning (HVAC) system, 44
HEPA filter, 30–31
High Efficiency Particulate Air (HEPA), 255
High risk deviation, 101–102
High risk elements, 75
Historical data set, 214
Human behavior, 12–13
Human fatigue, 45
Human performance, 12–13
Hydrogen peroxide gas, 256
Hypotheses testing, 223–224

I

ICH M4 Common Technical Document (CTD) format, 138
ICH Q8 Pharmaceutical Development Guidance, 138
ICH Q10 Pharmaceutical Quality System, 62
ICH Q9 Quality Risk Management, 62
Incubation, 50–51
Initial or short-term CPV, 212
In-Process Control Parameters (IPCs), 212
Inspection, labeling, and secondary packaging systems, 198
Installation qualification (IQ), 25, 188
Integrating risk management, 62
Integrity test, 233–234
Interbatch variability, 134
International Congress for Harmonization (ICH), 5
Intervention Risk Evaluation Method (I-REM), 47–49
Interventions, 46–47
  categories, 46–47
  evaluation and risk assessment methods, 47–50
  risk-based approach, 47
Intrabatch variability, 134
Isolators
  access, 256–257
  biological indicator locations, 259*t*
  Biological Indicators (BIs), 257–258
  challenges, 255
  chemical indicator locations, 258*t*
  decontamination process, 256
  design, 255
  Draeger tube locations, 259*t*
  equipment/components, 256
  hydrogen peroxide gas, 256
  throughput, 257
  validation, 258*t*, 259
  Vapor Phase Hydrogen Peroxide (VHP), 256–257

K

Key process parameters (KPP), 147, 162
Key quality management systems, 108
Keyword approach, 49, 73, 74*t*

L

Label coding, 198
Label reading systems, 198
Lethal treatment, 261–262
Likelihood of Occurrence Criteria, 73
Linear extrapolation, 159
Line of sight (LOS) approach, 17–20, 18*f*, 23, 69–70, 183–184
Long-term CPV, 213
Lower specification limit (LSL), 228
Low Risk, 75
  deviation, 100
Lyophilization, 198–199
Lyophilized product filling, 56–57

M

Mapping, aseptic process validation, 25–27
Master planning
  organization, and schedule planning, 139–141
  process validation activities and enablers/deliverables, 140*f*
Material storage, handling, and transport, 199
Measurement System Analysis (MSA), 133
Media fills, 10–11. *See also* Aseptic process simulations
  aborted and invalid, 55
  multiple, 40
  passed initially, 37–38
  run number, 39–41
Medium risks, 75
  deviation, 100–101
Microbial contamination, 35–38, 76–77
Microbiological behavior, 12–13
Microbiological contamination, 28–30, 53
  control, 180
Microorganism(s), 54
Mind maps, 75–76
Moist heat sterilization, 194, 261, 263
Monitoring, 183
  process variables, 212–213
Multivari chart, 151–152

N

The New Economics for Industry, Government, Education, 124–125
NOEL. *See* No observed effect level (NOEL)
Nonparametric data, 214

Nonparametric testing, 224
Nonsterility, nondestructive detection, 13
No observed effect level (NOEL), 84
Normality, 122–123
Normally distributed data, 214

O

Objective, defining, 133
Ointment filling, 56
One-factor-at-a-time, 152–153
Operational qualification (OQ), 25, 188–189
Outlier, 215
Out-of-specification (OOS), 215, 247
Out of trend, 215

P

Parametric data, 215
Parametric release, 266–267
Parenteral, 1
Parenteral Drug Association (PDA), 10, 51
Parenteral drug manufacturing, 26*f*
Parenteral process design
  control strategies and determine process design, 156–157
  design of experiments, 151–156
  master planning, organization, and schedule planning, 139–141
  pharmaceutical development, 137–138
  process/system design, 144–146
  risk assessment, performing, 146–151
  risk/impact assessment, 141–144
  scale-up and technology transfer, 157–162
  stage 1: design of experiments case study, 162–172
Parenteral process performance qualification
  cleaning validation, 200
  clean rooms and classified areas, 190–192
  component preparation, 192–193
  computer system validation, 200
  container closure integrity, 200–201
  critical utilities, 189–190
  equipment and systems, 189
  filling equipment and systems, 197–198
  inspection, labeling, and secondary packaging systems, 198
  line of sight approach, 183–184
  lyophilization, 198–199
  material storage, handling, and transport, 199
  number of batches, 202–205
  periodic assessment and requalification, 201
  principles, 177–183
  process qualification, 184–189
  product sterilization and filtration, 194–197
    dry heat sterilization, 195
    moist heat sterilization, 194
    radiation and ethylene oxide, 195
    sterile filtration, 195–197
  Stage 2b, 202
  terminal sterilization, 200
  testing laboratories, 199
Parenteral product manufacturers, 210–211
Parenteral product manufacturing, 9
Parts per million (PPM), 215
Patent or approval timing issues, 21
Performance indicators, 215
Performance metrics, 215
Performance qualification (PQ), 25, 189
Periodic assessment, 186
  and requalification, 27–28, 201
  schedule, 27
Periodic/time-driven requalification, 27
Permitted Daily Exposure (PDE), 83–84, 83*t*
Personnel (glove) microbial monitoring, 248*t*
Personnel Monitoring and Nonviables' data, 247
Pharmaceutical Development, 62
5-point scale, 73
Post aseptic fill lethal treatment, 265–266
Post aseptic fill sterilization, 262–265
Post aseptic processing sterilization, 262
Postincubation inspection, 53
Powder filling, 56
Power, statistical, 203–204
Preincubation inspection and rejection, 52–53
Preliminary Hazard Analysis (PHA), 71
Pressure decay, 233
Presterilization, 235
Preuse/poststerilization integrity testing, 234–236, 240*t*. *See also* Sterilizing grade filtration
  integrity test and downstream setup example for, 237*f*
  risks attached to, 236–241
Proactive and reactive culture, 98
Problem, defining, 133
Process Analytical Technology (PAT), 120, 157
Process capability, 14–15, 126–127, 134, 179, 228
  analysis, 231*f*
Process capability indices (PCIs), 122–123
Process control, 14–15, 34
Process design, 182, 215
Process life cycle approach, 16–17
Process map, 69–73

Process performance, 134
assurance, 17
Process Performance Indexes (Ppk), 126–127, 215
Process Performance Qualification (PPQ), 25, 89–90, 186, 202, 211–212, 215
Process qualification (PQ), 25, 63, 180, 182, 184–189, 215
Process requirements, 20–21, 161*t*
Process/system design, 144–146
Process understanding, 15–16
Process validation, 14–15, 179, 215–216
continuum, 5
CPV. *See* Continued process verification (CPV)
defined, 62, 210–211
life cycle approach, 94–96
parenteral process design. *See* Parenteral process design
parenteral process performance qualification. *See* Parenteral process performance qualification
requirements for drug products, 5
three-stage life cycle approach to, 16*f*
Process Validation: General Principles and Practices, 62
Process Validation Guidance for industry, 15
Product quality assurance, 14–15
Product quality reviews (PQR), 229–230
Product Specific Design Approach, 263
Product sterilization and filtration, 194–197
dry heat sterilization, 195
moist heat sterilization, 194
radiation and ethylene oxide, 195
sterile filtration, 195–197
Project charter, 21
Project execution timelines and budget, 98
Proportion, 216
Pseudomonas aeruginosa, 52
Pumping Number, 159
Reynolds number and, 159*f*
PUPSIT (poststerilization, preuse integrity testing) of sterilizing filters, 22
P-value, 91

Q

QMS. *See* Quality management system (QMS)
QRM. *See* Quality risk management (QRM)
Qualification, 24
Quality, 216
Quality Assurance (QA), 178, 217
Quality attributes
criticality, 147, 147*f*
criticality comparison and analysis, 149*t*
determination, 150*t*
scoring, 150*t*
Quality by design (QbD), 138, 152–153
Quality management system (QMS), 97–98, 99*f*, 109*t*, 217
Quality Risk Assessment Report, 162
Quality risk management (QRM)
assessment, scales for, 73–75
evolution, 61–64
fine print, 67
goal, scope and boundaries, 68–69
hazard, harm, and controls, 65–67, 66*f*
ICH Q9 and ICH Q10 definition, 64
ISO 14971 definition, 64
life cycle, 64, 65*f*, 75–79
risk assessment, 75–76
risk communication, 77–78
risk control, 76–77
risk review, 78–79
process map, 69–73
risk question and tool selection, 69–73
risk team membership, 67–68
Quality Target Product Profile (QTPP), 20–21, 141–142, 203

R

Radiation, 195
Randomization, 133
Rationales for Laboratory Characterization studies, 141
% Relative standard deviation (%RSD), 215
Replication, 133
Reynolds number, 157–159
equation, 159*f*
Pumping number and, 159*f*
Risk acceptance, 77, 143
decision, 145
Risk analysis, 75–76, 143
Risk assessment, 44, 68–69, 75–76, 146–151
criteria and definitions, 111*t*
template example, 112*t*
using compliance scoring criteria, 110
Risk-based criteria, 39
Risk-based decision making, 77
Risk-based life cycle approach, 82, 134
for process validation, 5
Risk-based prioritization, 99–100
Risk-based quality management system
change control, 105–106
corrective and preventive action, 102–105
deviation management system, 99–102
high risk deviation, 101–102

low risk deviation, 100
medium risk deviation, 100–101
risk assessment
criteria and definitions, 111*t*
template example, 112*t*
risk factors and scoring, 110–111
audit frequency using risk matrix, 112
audit schedule, 112–113, 112*t*
self-inspection process, 107–113
functional area categories, 109–110
key quality management systems, 108
risk assessment using compliance scoring criteria, 110
Risk-based self-inspection, 98
program, 108, 108*f*
Risk-based strategy, 62–63
Risk-based thinking, 62
Risk benefit analysis, 102–105, 103*t*
Risk category, 100
Risk communication, 77–78, 143
Risk control, 76–77, 143
and acceptance, 75*t*
Risk estimation matrix, 73–74, 75*f*
Risk evaluation, 143
Risk Facilitator, 67–68
Risk factors
ranking criteria, 50*t*
and scoring, 110–111
audit frequency using risk matrix, 112
audit schedule, 112–113, 112*t*
Risk identification, 143
Risk impact assessment, 141–144, 143*f*
Risk priority number (RPN), 73–74, 106–107
Risk question and tool selection, 69–73
Risk-ranking
criteria, 73
and filtering, 71
Risk reduction, 143
Risk registers, 77–78
Risk review, 78–79, 143
Risk scoring criteria, 103
Risk team membership, 67–68
Root Cause Analysis and plans, 99
Routine filtrative sterilization applications, 235–236
RPN. *See* Risk priority number (RPN)

S

Scale-up and technology transfer, 157–162, 158*t*
Secondary packaging systems, 198
Security systems, 199
Self-inspection process, 107–113
functional area categories, 109–110
key quality management systems, 108
risk assessment using compliance scoring criteria, 110
Semiautomated processes, 181
Semilog Survivor Curve Equation, 263
Severity multiplied by Occurrence, 148–149
Site acceptance testing (SAT), 25, 187–188
Special-cause variance, 216
Spore log reduction (SLR), 264
Stage 2a, 185–186
Stage 3a, 212
Stage 2b, 202, 218
Stage 3b, 212, 218
Standard sterility testing, 14–15
Staphylococcus aureus, 52
State of control, 216
Statistical capability analyses and run charts, 251
Statistical inference, 120
Statistical power, 203–204
Statistical Tolerance Intervals, 216
Statistical Tolerance Limit, 216
Factor, 216
Statistics in process validation
concepts, 120
control charting, 123–125
descriptive statistics, 123
design of experiments, 131–133
graphing of data, 120–121
hypothesis testing, 130–131
"let data talk to you,", 121
normality, 122–123
statistical inference, 120
tolerance and capability, 125–130
in parenteral manufacturing, 116, 117*f*
process and product specifications, 118–119
software packages, 135*t*
Steam damaged filter cartridge, 241*f*
Steam in place (SIP), 196
Steam sterilization qualification, 235
Sterile filtration, 195–197
Sterile vessel holding qualification, 35–38, 36*f*
Sterility assurance level (SAL), 35
Sterilization, 193, 262
Sterilized filtrate process equipment, 237–239
filter setup with associated pipe work, 238*f*
Sterilizers, 201
Sterilizing grade filtration. *See also* Preuse/poststerilization integrity testing
filtration risk assessment, 239*t*
integrity test employment ratio before enforcement, 234*t*

Sterilizing grade filtration (*Continued*)
preuse/poststerilization integrity, 240*t*
reasons to apply, 234–236
recommendations, 241–242
risks attached, 236–241
Support processes and systems, 35

T

Taguchi method, 151–152
Terminal sterilization, 9, 200
Testing, 183
laboratories, 199
Thermocouples, 194
Thermophilic biological indicators (Bis), 194
Things of value, 66
Timely user requirement specification, 22–23
Time ordered data analysis, 134
Tolerance and capability, 125–130
Tolerance interval, 125, 126*f*
Total Organic Carbon (TOC), 87–88
Traditional validation methods, 12–13
Train personnel, 35
Two one-sided *t*-test (TOST), 91–92

U

Uninterrupted HEPA filtered air, 30
Unqualified filling operation, 35
Upper Control Limits (UCL), 252, 252*f*
Upper specification limit (USL), 228
Upstream processing, 144–145
URS. *See* User requirement specifications (URS)
User requirement specifications (URS), 9, 22–23

V

Validation, 2–3, 14–15, 178*f*, 181
Vapor Phase Hydrogen Peroxide (VHP), 256
decontamination, 257
Variability, 90
inherent, 181
Variance, 216
Vectors of contamination, 12–13

W

"Worst-case" parameters, aseptic process simulation design, 42–43
Written records, CPV, 209–210

Printed in the United States
By Bookmasters